Evolution—Fact or Fiction?

Evolution—Fact or Fiction?

Nicholas Nurston

To order additional copies of this book, contact:
Xlibris Corporation
0-800-644-6988
www.Xlibrispublishing.co.uk
Orders@Xlibrispublishing.co.uk
304084

Darwin, Einstein and Newton made a search.

It is interesting to see a number of recent television programmes dedicated to 'Darwinian Evolution.' Although it is not surprising to hear its supporters, even on unrelated programme output, such as Stephen Fry for example, speak of Darwin's theories on natural selection as though proven fact. Darwin himself held greater misgivings about his theories than his present day advocates. His 'The Origin of Species' contains more than five thousand suppositions and yet not one fact in support of such theory. People such as David Attenborough and many other seemingly intelligent men of science, become quite ecstatic over the intricate design encountered in their respective fields of study, ignoring the fact that every design requires a designer.

So what facts do we have which suggest a great Creator of all things? Many of those seeking an answer from scripture have drawn a blank, including Darwin, who had to invent his own reason for the existence of matter and living creatures. Christendom let him down on the subject. Why, she cannot agree on, or understand what should be her instruction manual, the Bible. She has distorted the facts contained therein by introducing her own uncanny practices, festivals and rituals, thus succumbing to the many apostate forms of worship practiced before and since the advent of Christianity. She has removed the Devine name from all her literature and adopted the doctrines of men. Since the Nicaean Council of 325 C.E. she chooses not to accept the relationship between God and his Son and expects her subjects to believe, just as the Babylonians and ancient Greeks did, that upon death, an invisible being called the 'immortal soul' pops out of the body and flies off into the dwelling place of God.

Many ancient communities, although fiercely religious, searched for scientific fact and there is no doubt that Newton, a religious man, Einstein and others have been influenced by those discoveries, particularly in the field of Astrophysics. But where are these facts? In recent years it has been discovered that normal amounts of the blood-clotting element vitamin K are not found in an infant's blood until the fifth to seventh day after birth. Another clotting factor, prothrombin is present in amounts only about 30% of normal on the third day, but by the eighth day is higher than at any other time in a child's life, as much as 110% of normal. A wound at this stage would be better protected against haemorrhage. The Hebrew nation through Abraham, were instructed to carry out circumcision on the eighth day after birth. See Genesis 17:1, 9-14, 23-27. It was not necessary for the Grand Designer to explain to Abraham, as does Dr. S.I. McMillen to ourselves when he observes: "From a consideration of vitamin K and prothrombin determination, the perfect day to perform circumcision is the eighth day . . . [the day] picked by the Creator of vitamin K" See None of These Diseases, 1986, Page21.

2

SCIENCE AND SCRIPTURE:
ARE THEY COMPATIBLE?

More and more evidence is being uncovered to confirm the book of Isaiah was completed by one man at Jerusalem circa 732 B.C.E., although sceptics would have us believe it was written in the first century C.E., by several people.

Returning to scientific references contained in the Scriptures. Centuries before the shape of the Earth was indicated by Columbus and Magellan, Scripture stated, this planet is not flat but spherical, Isaiah 40:22. Long before astronauts had a glimpse of the earth hanging in empty space, the Bible pointed out that it is suspended upon nothing. Job 26:7. Both Newton and Einstein appreciating grand design, wrestled with the understanding of a Master Designer, though had he but known it, Einstein could have turned to Isaiah 40:26 which speaking of an abundance of dynamic energy, perfectly describes his $E=mc^2$ which will be considered in the concluding Genesis argument. In 1910 The Encyclopaedia Britannica explained: "matter can neither be created nor destroyed." Back then that would seem to be a reasonable assumption. However in 1905 Einstein, as stated, predicted the relationship between matter and energy which led other scientists to explain how the sun has kept shining for billions of years. Today, scientists accomplish the conversion of energy to matter by use of particle accelerators. "We're repeating one of the miracles of the universe transforming energy into matter," explains Nobel laureate physicist Dr. Carlo Rubbia.

There are so many such examples contained in the scriptures such as earth's water cycle which we are only just beginning to understand. Some 3000 years ago, the Bible described in simple terms, the earth's water cycle as part of the ecosystem which makes life possible. Ecclesiastes 1:7. The point to be made is that both the Hebrew and Greek scriptures, commonly known as the Old and New Testaments, were never intended to be a treatise on the design and scientific complexities of universal creation. Were they to be so, would we at any stage of learning, be capable of understanding? See Isaiah 40:28. Nevertheless, to those not blinded by prejudice and able to use the power of reasoning, the instances recorded in simple terms, provide a far more plausible explanation than Darwin's hit or miss ideas of chance. Were he living today and able to draw on all the advances of our current accumulated scientific knowledge, it is likely he would be first to scoff at his own ideas. Unfortunately, it is also likely that his present day disciples would still adhere to the outdated notions he carried, happy in the belief that at some obscure point in time, they had lost their tails.

King David was inspired to record his awe when considering the human body, stating that "in a fear-inspiring way I am wonderfully made." He goes on to consider his embryo and refers to the book or blueprint containing its genetic make-up. That information imparted by a man who could never have had knowledge of genetics from his own understanding. Psalm: 139: 13-16. The bible is greatly concerned with the issue of justice, sanctification, vindication and the fulfilment of prophecy, all of which have proven to be very accurate to this day and in particular, the current unprecedented world social and economic conditions. 2 Timothy 3:1-5.

RADIO CARBON DATING:
HOW ACCURATE?

A common misunderstanding is the period of man's existence. The only recorded history dates back a mere 6,500 years. Some, without evidence, claim humankind has existed for millions of years, give or take a million or so. Even planet Earth, although dating back millions of years, is a relative newcomer to the One having no beginning or end and yet being the Alpha and Omega of all things Revelation 1:8, 21:6, 22:13. Use of Radio Carbon Dating assumes the sun's effect upon the Earth has been constant in the degree and type of radiation throughout earth's history. Could though, an occurrence of great magnitude have changed that condition? It is at this point that we should consider the feasibility of the Biblical Flood account and its possible effect on Earth's climate and geological form. The Bible speaks of a volume of water above the expanse, which together with springs, produced the cataclysmic Deluge of 2370 B.C. Anthropologists confirm, with that deluge great changes came about. Earth's population and the life-span of humans dropped very rapidly. Some have suggested that prior to the Flood, the Waters above the expanse shielded out some of the harmful radiation and with that protective liquid canopy no longer in place, cosmic radiation, genetically harmful to man, increased. However, the Bible is silent on the matter. On returning to radio carbon dating, we might consider that any change in radiation would have altered the formation rate of radioactive carbon 14, to an extent which would invalidate all radio carbon deterioration date calculation prior to that Flood. With the sudden opening of the 'springs of the watery deep' and "the floodgates of the heavens," untold billions of tons of water deluged the Earth. [Gen. 7:11] This may have caused tremendous changes in Earth's surface. Its crust being relatively thin, estimated at between30 km and 160 km thick, stretched over a somewhat plastic mass, thousands of miles in diameter. Under that added weight of water, there was likely a great shifting in the crust. In time new mountains evidently were thrust upward, old mountains rose to new heights, shallow sea basins were deepened and new shorelines established. The result is that now about 71% of the surface is covered by seawater. This shifting in Earth's crust may account for many geological phenomena, such as the raising of old coastlines to new heights. It has been estimated by some, that water pressures alone, would have been equal to "two tons per square inch," sufficient to fossilize fauna and flora quickly. See Bible Flood and the Ice Epoch, by D. Patten 1966 p. 62. It is a fact that during Autumn, leaf-fall causes the rotation of the Earth to speed up as the result of a weight shift closer to Earth's centre of gravity. In comparison, we may wonder what difference in rotational speed such a massive weight of water would have produced. A number of theories have been put forward to account for how such a massive amount of water could be held in suspension above "The Expanse". Giving it thought, was it held in position by pressure differential, with a full or partial vacuum above the spatial layer reducing the effect of gravity on the water and atmospheric pressure below that layer producing equilibrium?

We can only guess, but it would not be difficult to achieve by the One, able to produce all the matter in the universe as the result of "an abundance of dynamic energy." After all it would have been this same energy which brought about the imbalance, producing the Deluge. We must also remind ourselves, before that Deluge, the Earth would have enjoyed an atmosphere similar to a giant greenhouse, producing a uniform temperature from pole to pole and probably the real reason why mammoth with edible food in its mouth, an indication that its death was sudden, has been discovered in what are now deep areas of permafrost.

Some would have us believe its extinction was due to a meteorite hitting Earth in the tropical region of America and causing a cloud which blanked the sun's rays. If that were so, it would have taken a period of time to produce death and it is unlikely food would have been found in an animal's mouth. Equatorial plants have also been discovered in such areas, lending weight to the theory that temperature throughout the earth was uniform. People sometimes scoff at the story of the 'Rainbow Covenant' when Noah and his family were told after the deluge, that as long as he was able to see a rainbow, mankind would never again be destroyed by that means. Under greenhouse conditions Earth would have been watered by mist (Genesis 2:6) and the sun's rays would not have been able to shine through rain droplets because there were none. After the watery canopy was poured out upon the earth and the flood waters receded, then under normal shower conditions a rainbow has since been seen regularly. It was at that time the polar caps would have formed, extreme differentials in temperature would have been experienced and turbulent winds would have arisen, disturbing what hitherto had been a still and tranquil atmosphere without inter-tropical convergence zones which produce our hurricanes and typhoons. The destructive energy contained in water powered by tsunami has been brought home to us in recent years and that would be a mere dribble compared to a world covering deluge. A tsunami wave usually approaches land in one direction only, the movement of the currents in a worldwide deluge would head in all points of the compass and would be accountable for massive boulders lying in the most unusual places, a phenomenon geologists explain as the result of glaciers, but overlooking the fact that glaciers too, only move in one direction.

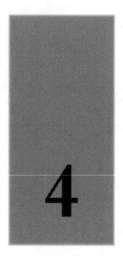

COSMIC EXPLOSION OR DESIGN?

Now let us take a brief mental trip into space to see earth and sun in perspective. Our sun, which could hold a million earths inside it with room to spare, Is just one of an awesome number of stars in a spiral arm of the Milky Way Galaxy which itself is some 600 quadrillion miles in diameter, taking light 100,000 years to cross it. Although containing over 100 billion stars, it is just a tiny part of the universe. The Milky Way, Andromeda and some twenty other galaxies are bound gravitationally into a cluster, all of which are a small neighbourhood in a vast super-cluster. The universe contains countless super-clusters and the picture does not end there because they are not evenly distributed in space. On a grand scale, they look like thin sheets and filaments around vast bubble-like voids, some features of which are so long they resemble great walls.

This may surprise many who think our Universe created itself in a chance cosmic explosion. "The more clearly we can see the universe in all its glorious detail" concludes a senior writer for 'Scientific American,' "the more difficult it will be for us to explain with simple theory how it came to be that way." All the individual stars we see are in the Milky Way Galaxy! Until the 1920's that seemed to be the only galaxy. Observations with larger telescopes of course, have since proven otherwise. Our universe contains at least fifty billion galaxies, each containing billions of stars like our sun. What shook scientific beliefs in the 1920's was not the staggering quantity of huge galaxies, but the fact that they were all in motion. Astronomers went on to discover a remarkable fact when galactic light was passed through a prism, the light waves were seen to have been stretched, indicating motion away from us at great speed. The more distant a galaxy, the faster it appeared to be receding, pointing to an expanding universe. In 1995 this was confirmed by observation of the most distant star (SN 1995K) ever observed as it exploded. New Scientist magazine plotted this on a graph and explained: "the shape of the light curve . . . is stretched in time by exactly the amount expected if the galaxy was receding from us at nearly half the speed of light." Conclusion, this is the "best evidence yet that the Universe really is expanding." Even if we are neither, professional or amateur astronomers, we are able to see that an expanding universe would hold profound implications with regard to our past and perhaps our future too. Something must have started the process, a force powerful enough to overcome the immense gravitational forces of the entire universe. We all have good reason to ask, 'what could be the source of such dynamic energy?'

5

Fundamental Forces—
Gravity and Electromagnetism.

Although most scientists trace the universe back to a very small,
dense beginning (a singularity), we cannot avoid the issue: "If
at some point in the past, the Universe was close to a singular
state of infinitely small size and infinite density, we have to ask what was
there before and what was outside the Universe . . . We have to face the
problem of a beginning"—Sir Bernard Lovell. This implies more than
just a source of vast energy. Foresight and intelligence are also required,
because the rate of expansion seems to be very finely tuned. "If the
Universe had expanded one million millionth part faster" said Lovell,
"then all the material in the Universe would have dispersed by now . . .
And if it had been a million millionth part slower, gravitational forces
would have caused the Universe to collapse within the first thousand
million years or so of its existence. Again, there would have been no
long-lived stars and no life." Can experts, now convincingly explain the
origin of the Universe?

Many scientists, uncomfortable with the idea that it was created by a
higher intelligence, speculate that by some mechanism it created itself out

of nothing. Does that sound reasonable? Such speculations usually involve a variation of a theory (inflationary universe model) conceived in 1979 by physicist Alun Guth. More recently, Dr. Guth admitted that his theory "does not explain how the Universe arose from nothing." Dr. Andre Linde was more explicit in a 'Scientific American' article: "Explaining this initial singularity—where and when it all began—still remains the most intractable problem of modern cosmology."

Experts cannot really explain either the origin or early development of our Universe and therefore we have valid reason to consider evidence which many have overlooked and should give us real insight into the issue. The evidence involves the precise measurement of four fundamental forces which are rarely if ever referred to by those presenting TV programmes and claiming to explain the formation of Earth and the emergence of life upon it. These four fundamental forces are [**Gravity, Electromagnetism, Strong nuclear force and Weak nuclear force**] and are responsible for all properties and changes affecting matter. Because they affect us personally, the basic facts are worthy of consideration.

The four fundamental forces come into play both in the vastness of the cosmos and in the infinite smallness of atomic structures. In fact, everything we see around us. Elements vital for life, (particularly carbon, oxygen and iron) could not exist were it not for the fine tuning of the four forces evident in the Universe. One force, gravity has already been mentioned, another is electromagnetic force. If it were significantly weaker, electrons would not be held around the nucleus of an atom and therefore, atoms would be unable to combine to form molecules. Conversely, if this force were much stronger, electrons would be trapped on the nucleus of an atom. There could be no chemical reactions between atoms and subsequently, no life. Even from this standpoint, it is evident that our existence and life, are dependent upon the fine-tuning of the electromagnetic force. Consider in the cosmic scale: A slight difference in the electromagnetic force would affect the sun and thus alter the light reaching Earth, making photosynthesis difficult or impossible. It could rob water of its unique properties, vital for life. Again, the precise tuning of the electromagnetic force determines whether or not we live. The relationship of the electromagnetic force is equally vital to the other three forces. Some physicists calculate this force to be 10^{40} times that of gravity.

It may seem a small change on that figure to add one more zero 10^{41} and yet it would result in gravity being proportionally weaker. Dr. Reinhard Breur comments on the resulting situation: "With lower gravity the stars would be smaller and the pressure of gravity in their interiors would not drive the temperature high enough for nuclear fusion reactions to get under way: the sun would be unable to shine." If gravity became proportionately stronger so that the number of zeros were decreased by one (10^{39}): "With just this tiny adjustment," continues Breur, "a star like the sun would find its life sharply reduced." Yet other scientists consider this fine-tuning to be even more precise. Two remarkable qualities of our sun and indeed other stars are long-term stability and efficiency. Automotive engineers are always striving to improve the performance of the internal combustion engine by optimizing the fuel/air mixture to produce the most efficient combustion. In our sun the key forces involved are precisely tuned to sustain life. Did that just happen by chance? The ancient man Job was humbled when asked: "Did you proclaim the rules that govern the heavens, or determine the laws of nature on earth?" (Job 38:33 The New English Bible) No human did. So from where does the precision originate?

FUNDAMENTAL FORCES—
STRONG AND WEAK NUCLEAR FORCES.

The structure of the universe involves much more than fine-tuning just gravity and the electromagnetic force. Two other physical forces also relate to our life. These operating in the nucleus of an atom give ample evidence of forethought. The **strong nuclear force** binding protons and neutrons in the nucleus, permitting various elements to form—light ones such as helium and oxygen and heavy ones such as gold and lead. It would seem if that binding force were 2% weaker, only hydrogen would exist. Conversely, if that force were slightly stronger, only heavier elements, but no hydrogen could be found. This would of course produce a catastrophic effect upon our existence. Our sun would not have the fuel it requires to radiate life-giving energy. We would have no food or water, since hydrogen is an essential ingredient of both. From—'New Scientist'—headed "Combination of Coincidences"—"Make the weak force slightly stronger and no helium would have been produced; make it slightly weaker and nearly all the hydrogen would have been converted to helium."—"The window of opportunity for a universe in which there is some helium and there is also exploding supernovas is very narrow. Our existence depends on this combination of coincidences and on the even more dramatic coincidence of nuclear energy levels predicted by Hoyle. Unlike previous generations, we know how we came to be here. But, like all previous generations we do not know why."

The forth force, **weak nuclear force**, controls the rate of radio-active decay. It also has an effect upon the thermonuclear activity in our sun. Mathematician and physicist Freeman Dyson explains: "The weak [force] is millions of times weaker than the nuclear force. It is just weak enough so that the hydrogen in the sun burns at a slow and steady rate. If the weak [force] were much stronger or much weaker, any forms of life dependent on sun-like stars would again, be in difficulties." That precise rate of combustion keeps our earth warm—but not incinerated, thus keeping us alive. Furthermore, scientists believe the weak force plays a role in supernova explosions, which they give as the mechanism for producing and distributing most elements, "If those nuclear forces were in any way slightly different from the way they actually are, the stars would be incapable of making the elements of which you and I are composed," explains physicist John Polkinhorne. More could be said, but the point has been made that there is an amazing degree of fine-tuning of these four fundamental forces. "All round us, we seem to see evidence that nature got it just right," wrote Professor Paul Davies. The precise tuning of the fundamental forces has made possible the existence and operation of our sun, our delightful planet with its life-sustaining water, our atmosphere so vital for life and a vast array of precious chemical elements on earth. This must compel us to ask, Why such precise tuning and from where?

DARWIN TRIED TO LOOK
BEYOND RELIGIOUS MYTHS.

A certain amount of credit must be given to Darwin, because he was making an attempt to see beyond many of the religious myths, untruths, explanations and misguided practices upon which he had been weaned.

His enthusiasm was genuine in his quest for truthful answers, but one can be genuinely wrong. Remember, he did not have the advantage of the knowledge or understanding available to us. Had he been so informed, as suggested earlier, I am sure he would have been quite embarrassed by his reasoning and suppositions.

With regard to Earth's ideal features: let us consider its measurements and position relative to the rest of our solar system. The Bible book of Job contains more humbling questions: "Where did you happen to be when I founded the earth? . . . Who set its measurements, in case you know?" (Job38:4-5) As never before, those questions beg for answers. Why? : Because of the amazing things which have been discovered since Darwin's day. No planet like Earth has been found elsewhere in the Universe. True, some scientists point to indirect evidence that certain stars have orbiting them; objects hundreds of times larger than our earth. If we consider its measurements and position relative to the rest of our solar system: Earth is just the right size to support our existence. If it were slightly larger, its gravity would be stronger and hydrogen would collect, unable to escape Earth's gravity and consequently, it would be inhospitable to life. If on the other hand it was slightly smaller, then life-sustaining oxygen would escape and surface water would evaporate. Under either of those two conditions, we could not live! "The special conditions on earth resulting from its ideal size, element composition and nearly circular orbit at a perfect distance from the sun, made the accumulation of water on the earth's surface." [Integrated Principles of Zoology. 7th edition] Earth is also positioned an ideal distance from the sun, another factor vital for life to thrive. Astronomer John Barrow and mathematician Frank Tipler studied "the ratio of the Earth's radius and distance from the Sun." They concluded; human life would not exist "Were this ratio slightly different from what it is observed to be." Professor David L. Block notes: "Calculations show that had the earth been situated only 5% closer to the sun, a runaway greenhouse effect [overheating of the earth] would have occurred about 4000 million years ago. If on the other hand, the earth were placed only 1% further from the sun, runaway glaciations [huge sheets of ice covering much of the globe] would have occurred some 2000 million years ago."—Our Universe: Accident or Design? To the evidence of the foregoing precision we can add the fact that the earth rotating once a day on its axis; does so at a speed which produces moderate temperatures. If like Venus, it took 243 of our days to complete one revolution, we would be unable to survive the extreme temperatures resulting from such long days and nights. Vital too is our earth's path around the sun. Comets follow a wide elliptical path. Thankfully this is not so with the earth. Its orbit, being almost circular, prevents us from experiencing death-dealing extremes of temperature. Furthermore we cannot ignore the location of our solar system. Were it nearer the centre of the Milky Way galaxy, the gravitational effect of neighbouring stars would distort that orbit. Contrastingly, were it situated at the very edge of our galaxy; the night sky would be all but devoid of stars. Starlight, although not essential to life, does add great beauty to the night—sky.

WHAT DO RESPECTED
SCIENTISTS HAVE TO SAY?

B ased upon current concepts of the universe, scientists calculate that at the edges of the Milky Way, there would not have been enough chemical elements to form a solar system like ours. These elements, scientists have found reveal amazing order and harmony. A modern encyclopaedia of science describes the chemical elements as "Architectural Units of the Universe." How the atoms of those elements are constructed and relate to each other; bespeaks economy and awesome organization in chart-like order. Around three hundred years ago, only—antimony, arsenic, bismuth, carbon, copper, gold, iron, lead, mercury, silver, sulphur and tin were known to man. As more were discovered, scientists noticed the elements reflected a distinct order. Because there were gaps in the order, scientists such as Mendeleyev, Ramsay, Moseley and Bohr theorized the existence of unknown elements and their characteristics. Those elements were subsequently discovered as predicted. This could be achieved because all elements follow a natural numerical order based upon the structure of their atoms: That proven law thus enabling school textbooks to set out a periodic table of elements. The 'McGraw-hill Encyclopaedia of Science & Technology' observes: "Few systemizations in the history of science can rival the periodic concept as a broad revelation of the order of the physical world; . . . Whatever new elements may be discovered in the future, it is certain they will find a place in the periodic system, conforming to its order and exhibiting the proper familial characteristics." Do they place themselves in order?

Other men of science who offer a view on the subject—Robert Jastrow, Professor of Astronomy and Geology at Columbia University, wrote: "Few astronomers could have anticipated that this event—the sudden birth of the Universe—would have become a proven scientific fact, but observations of the heavens through telescopes have forced them to that conclusion." On the implications, he commented: "The astronomical proof of a beginning places scientists in an awkward position; for they believe that every effect has a natural cause." The British astronomer E.A. Milne wrote, "We can make no propositions about the state of affairs [in the beginning]; in the Divine act of creation God is unobserved and unwitnessed." 'The Enchanted Loom—Mind in the Universe.' Sir Fred Hoyle explains in the 'The Nature of the Universe': "To avoid the issue of creation it would be necessary for all the material of the Universe to be infinitely old and this cannot be. Hydrogen is being steadily converted into helium and the other elements. How comes it then that the Universe consists almost entirely of hydrogen? If matter were infinitely old this would be quite impossible. So we see that the Universe being what it is, the creation issue cannot be dodged."

WHAT IS THE TREE OF LIFE?

We have discussed the formation of matter and the Universe and considered quoted extracts from the thoughts and findings of experts in their particular fields by citing just a few factual discoveries. It is now appropriate to consider Darwin's theory and view his findings on natural selection and the 'Tree of Life.'

I have particularly chosen the latter because David Attenborough has bluffed his way through an hour-long BBC television programme of that title. Confirming his adoration of Darwin, he tossed in a few scientific facts in the hope that they might bamboozle people into accepting that living organisms originated by chance from a primordial soup. As we examine the facts he touched upon, it will become evident that he used them only vaguely and completely out of context. It would seem that to the producers of such programmes, the factual content is probably not that important providing it entertains and follows the BBC policy of choosing evolution as the only reason for the existence of life. The latter confirmed in letter form by Head of Science, BBC Factual who, when presented with the scientific facts contained in this document stated "I can only respond that the BBC's clear position is that evolution is a scientific theory. Scientific theories are established by being tested against factual evidence." Could not agree more, show us the factual evidence please! BBC is shirking its responsibility as an educator, by broadcasting fictitious fairy tales disguised as fact. The loveable David Attenborough has for years been responsible for bringing the wonders of nature to our screens. However, in doing so, he has constantly stuck his head in the sand and preached his Darwinian ideas, oblivious to the massif strides which have been made in his own field and what is more, the responsibility he carries. That responsibility is to all, but particularly to young people, who often carry those teachings through their primary education and mostly holding on to them for life, in the belief that what they were taught was factual. BBC Factual is a misnomer and perhaps should be changed to 'BBC Almost Factual.'

WHAT IS CHANCE?

From the foregoing, it is almost certain that a good deal of sympathy and agreement would be given by BBC to the statement made by Nobel laureate Christian de Duve, who speaking about the origin of life said: "Chance and chance alone, did it all, from the primeval soup to man." So by analyses then, firstly what is chance? Some think in terms of a mathematical probability, such as that involved in flipping a coin. That though, is not how many scientists use 'chance' regarding the origin of life. The vague word 'chance' is used as a substitute for a more precise word such as "cause," especially when the "cause" is unknown. "To personify 'chance' as if we were talking about a casual agent," notes biophysicist Donald Mackay, "is to make an illegitimate switch from a scientific to a quasi-religious mythological concept." Similarly, Robert C. Sproul points out: "By calling the unknown cause 'chance' for so long, people begin to forget that a substitution was made . . . the assumption that 'chance equals an unknown cause' has come to mean for many that 'chance equals cause'." Others, who reasoned in this fashion, Nobel laureate Jacques L. Monod, for one, used this chance-equals-cause line of reasoning. "Pure chance, absolutely free but blind, [is] at the root of the stupendous edifice of evolution," he wrote. "Man knows at last he is alone in the universe's unfeeling immensity, out of which he emerged only by chance."

Note he says: 'By chance.' Monod does what David Attenborough has done—he elevates chance to a creative principle. Chance is offered as the means by which life came to be on earth. Dictionaries show that "chance" is the assumed impersonal purposeless determiner of unaccountable happenings." "Chance"; written with a capital letter—in effect saying, Creator?

DNA and RNA—
Awesome Interaction!

We are all aware our earth teems with life. From the snowy Arctic to the Amazon rain forest, from the Sahara Desert to the Everglades swamp, from dark ocean floor to bright mountain peaks—life abounds. And it is loaded with the potential to amaze us. It comes in types, sizes and quantities which stagger our imagination: A million species of insects, 20,000 species of fish—some the size of a grain of rice, others as long as a London bus. At least 350,000 plant species—some weird, most wonderful—embellish the land. Over 9,000 species of birds fly overhead. All these creatures, including man, form the panorama and symphony that we refer to as life. More amazing than the delightful variety around us, is the profound unity which links them. Biochemists, who look beneath the skin of earth's creatures, explain that all living things—be they amoebas or humans—depend upon an awesome interaction: the teamwork between nucleic acids (DNA and RNA) and protein molecules. The intricate process involving those components occur in virtually all our body cells, as it does in the cells of all living things, producing as it does, a beautiful mosaic of life. It is generally accepted that Earth at one time had no life upon it. Scientific opinion agrees, and so do many religious books. Both those sources though—science and religion—differ in explaining how life on earth began. Millions of people of all educational levels believe an intelligent Creator, the original Designer, produced life here on earth. In contrast, many scientists say life arose from non-living matter, one chemical step after another, merely by chance. Is it one, or is it the other? As already noted, one of the very fundamental questions humans have sought to answer is: Where do we, as living humans come from? Most science courses focus on the adaption and survival of life-forms, rather than on the more central question, 'the very origin of life.' As we have seen with David Attenborough, attempts to explain where life came from are often presented in generalizations such as: 'Over millions of years, molecules in collision somehow produced life.' Is that answer really satisfying? Because that would mean that in the presence of energy from the sun, lightning or volcanoes, some lifeless matter moved, became organized, and eventually started living—all without direct assistance. What a huge leap: 'From non-living matter to living.' Bearing in mind the comments on 'chance,' I ask the reader not to make a judgment on the foregoing until consideration of the following submissions with regard to experiments aimed at proving that claim. Back in the middle ages, the concept of a huge leap from non-living to living matter would have been accepted, because spontaneous generation was a prevailing belief.

Finally in the 17th century, Italian physician Francesco Redi, proved that maggots appeared in rotten meat, only after flies had laid eggs on it. No maggots developed on meat that flies could not reach. If animals as big as flies did not just appear on their own, what about the microbes that kept appearing in food—covered or not? Although later experiments indicated that microbes did not arise spontaneously, the issue remained controversial until work by Louis Pasteur. Although known primarily through his work on infectious disease, Pasteur performed experiments to determine if tiny lif-forms could arise by themselves. He demonstrated that even minute bacteria did not form in sterilized water protected from secondary contamination. In 1864 he announced: "Never will the doctrine of spontaneous generation recover from the mortal blow struck by this simple experiment." That statement of course, remains true to this day. No experiment has ever produced life from non-living matter. Once again we must ask: How then could life come to be on Earth? In spite of this fact, in the 1920's Russian biochemist Alexander I. Oparin and other scientists since, have offered something like the script of a three-act drama depicting what is claimed to have occurred on the stage of planet Earth. The first act portrays earth's elements, or raw materials, being transformed into groups of molecules. In the second act: The jump to large molecules. And the last act of this drama presents the leap to the first living cell. Now fundamental to that drama is a claim that earth's early atmosphere was much different to that of today. One theory and I stress, it is only a theory and not fact, assumes that free oxygen was virtually absent and the elements nitrogen, hydrogen and carbon, formed ammonia and methane. The concept being, that when lighting and ultraviolet light struck an atmosphere comprised of those gasses and water vapour, sugars and amino acids, developed. According to this theoretical drama, such molecular forms washed into the oceans or other bodies of water. Over time, sugars, acids and other compounds concentrated into a broth of "pre-biotic soup" where amino acids, for instance, joined to become proteins. Extending this theoretical progression, other compounds called nucleotides formed chains and became a nucleic acid such as DNA. All of this supposedly set the stage for the final act of the molecular drama. One might depict this last act, which is undocumented, as a love story. Protein molecules and DNA molecules happen to meet, recognize each other and embrace. Then just before the curtain rings down, the first living cell is born. Following that dramatization, it may be wondered; Is this real life or fiction? Could life on earth have originated that way?

Attempts to create life—
An ongoing frustration.

I n the early 1950's, Oparin's theory was tested. It was an established
fact that life comes only from life and yet scientists theorized that
if conditions differed in the past, life may have come slowly from
non–life. Hoping to demonstrate this, Stanley L, Miller, working in the
laboratory of Harold Urey, took hydrogen, ammonia, methane and water
vapour (assuming that to have been the primitive atmosphere), sealed
these in a flask with boiling water at the bottom (representing the ocean),
passed a high voltage electric current through the vapours (resultant
sparks representing lightning).

Within a week, there appeared traces of reddish goo; Which Miller
analyzed to discover it to be rich in amino acids which are of course
the essence of proteins. Until this day, that experiment has been cited in
science textbooks and school courses as if explaining the origin of life.
Millers experiment of 1953 though, is seriously questioned today but
often cited as evidence that spontaneous generation could have occurred
in the past. The validity of his explanation, rests on the presumption that
the earth's primordial atmosphere was "reducing" meaning, it contained

only the smallest amount of free (chemically un-combined) oxygen. Now why is that?

'The Mystery of Life's Origin': 'Reassessing Current Theories' points out that if much free oxygen was present, "none of the amino acids could even be formed and if by some chance they were, they would, (because of the highly reactive nature of oxygen), decompose quickly." How solid was Miller's presumption about the so-called primitive atmosphere? In a classic paper published two years after his experiment he wrote: "These ideas are of course speculation, for we do not know that Earth had a reducing atmosphere when it was formed . . . No direct evidence has yet been found."—Journal of the American Chemical Society, 12th May, 1955.

Was evidence ever found? Some 25 years later, science writer Robert C. Cowen reported: "Scientists are having, to rethink some of their assumptions . . . Little evidence has emerged to support the notion of a hydrogen-rich, highly reducing atmosphere, but some evidence speaks against it." Technology Review, April 1981. Since that date, ten years later in 1991, John Horgan wrote in Scientific American: "Over the past decade or so, doubts have grown about Urey and Miller's assumptions regarding the atmosphere. Laboratory experiments and computerized reconstructions of the atmosphere . . . suggest that ultraviolet radiation from the sun, which today is blocked by atmospheric ozone, would have destroyed hydrogen-based molecules in the atmosphere . . . Such an atmosphere [carbon dioxide and nitrogen] Would not have been conducive to synthesis of amino acids and other precursors of life."

Science textbooks—
Premature findings?

Despite the overwhelming volume of evidence against it many like David Attenborough still hold that earth's early atmosphere was reducing, containing little oxygen! In 'Molecular Evolution and the Origin of Life,' Sidney W. Fox and Klaus Dose claim the atmosphere must have lacked oxygen because, for one thing, "laboratory experiments show that chemical evolution ... would be largely inhibited by oxygen" and because compounds such as amino acids "are not stable over geological time in the presence of oxygen." Is this not circular reasoning? The early atmosphere was a reducing one, it is said, because spontaneous generation of life could not otherwise have taken place. But there is no assurance that it was reducing! There is another telling detail: If the gas mixture represents the atmosphere, the electric spark mimics lightning and boiling water stands for the sea, what or who does the scientist, arranging and carrying out the experiment represent?

Throughout the passage of time all that optimism has evaporated. Professor Miller, some forty years after his experiment, told 'Scientific American:' "The problem of the origin of life has turned out to be much more difficult than I and most other people envisioned." Other scientists share his change of thought. One example being Professor of Biology Dean H. Kenyon, who back in 1969 co-authored 'Biochemical Predestination.' More recently he concluded that it is "fundamentally implausible that unassisted matter and energy organized themselves into living systems." Indeed, laboratory work bears out Kenyon's assessment that there is "a fundamental flaw in all current theories of the chemical origins of life." After Miller and others had synthesized amino acids, scientists set out to make proteins and DNA, both of which are undisputedly necessary for life on earth. So then, after thousands of experiments with so-called pre-biotic conditions; what was the outcome? 'The Mystery of Life's Origin: Reassessing Current Theories' quoted from in an earlier paragraph, notes: "There is an impressive contrast between the considerable success in synthesizing amino acids and the consistent failure to synthesize protein and DNA." The latter efforts are characterized by "uniform failure." Realistically, the mystery encompasses more than how the first protein and nucleic acid (RNA or DNA) molecules came into existence. It includes; how do they work together? "It is only the partnership of the two molecules that makes contemporary life on Earth possible," says 'The New Encyclopaedia Britannica,' going on to note, how that partnership could come about remains "a critical and unsolved problem in the origin of life." Truth indeed!

BEND THE FACTS TO SOLVE
UNSOLVABLE RIDDLE.

Professor of Biology, Lyn Margulis stated—"[The smallest bacterium] is so much more like people than Stanley Miller's mixtures of chemicals, because it already has these system properties. So to go from bacterium to people is less of a step than to go from a mixture of amino acids to that of bacterium." Taking this a step further, we know that there are right—handed and left-handed amino acid molecules. Of some one hundred known amino acids, only twenty are used in proteins, all of which are of left-handed type. When scientists in laboratories make amino acids in imitation of what they feel possibly occurred in a pre-biotic soup, they find an equal number of right-handed and left-handed molecules. "This kind of 50/50 distribution," reports 'The New York Times,' is "not characteristic of life, which depends on left-handed amino acids alone." Dr. Jeffery L. Balda, who studies problems involving the origin of life, said "some influence outside the earth might have played some role in determining the handedness of biological amino acids." A glimpse into the realm of our body cells elicits admiration for the work of scientists in this field. They have shed light on extraordinary complex processes that few of us even think about, although they operate at every moment of our lives.

Origin-of-life scientists have not ceased trying to formulate a plausible scenario for the drama about the first appearance of life and their new scripts are not proving to be any more convincing. For example, Klaus Dose of the Institute of Biochemistry, Mainz, Germany, observed: "At present all discussions on principle theories and experiments in the field either end in stalemate or in a confession of ignorance." At the 1996 International Conference on the Origin of Life, no solutions were forthcoming. The journal 'Science' reported that nearly 300 scientists who attended had "grappled with the riddle of how [DNA and RNA] molecules first appeared and how they evolved into self-reproducing cells."

British astronomer Sir Fred Hoyle has spent decades studying the universe and life in it, to the extent that at one time suggesting life on earth arrived from outer space. Lecturing at the California Institute of Technology he discussed the order of amino acids in proteins. "The big problem in biology," said Hoyle, "isn't so much the rather crude fact that protein consists of a chain of amino acids linked together in a certain way, but that the explicit ordering of the amino acids endows the chain with remarkable properties . . . if amino acids were linked at random, there would be a vast number of arrangements that would be useless in serving the purpose of a living cell. When you consider that a living enzyme has a chain of perhaps 200 links and that there are 20 possibilities for each link, it's easy to see that the number of useless arrangements is enormous, more than the number of atoms in all the galaxies visible in the largest telescopes. This is for one enzyme and there are upwards of 2000 of them, mainly serving very different purposes. So how did the situation get to where we find it to be?" Hoyle added: "Rather than accept the fantastically small probability of life having arisen through the blind forces of nature, it seemed better to suppose that the origin of life was a deliberate act." After nearly a half century of speculation and thousands of attempts to prove life originated on its own by chance, it would be hard to disagree with Nobel laureate Francis Crick, who speaking about origin-of-life theories observed, there is "too much running after too few facts."

Advanced science cannot prove that life could arise by itself and yet some scientists continue to hold such theories! A few decades ago, Professor J.D Bernal offered some insight in the book, The Origin of Life: "By applying the strict cannons of scientific method to this subject [the spontaneous generation of life] it is possible to demonstrate effectively at several places in the story, how life could **not** have arisen; the improbabilities are too great, the chances of the emergence of life too small." He added: "Regrettably from this point of view, life is here on Earth in all its multiplicity of forms and activities and the arguments have to be bent round to support its existence." Our considering the underlying import of such reasoning must give rise to scepticism, because it is saying: Scientifically it is correct to state life cannot have begun by itself. But spontaneously arising life is the only possibility we will consider. Therefore to support the hypothesis that life arose spontaneously it is necessary to bend the facts.

THE RNA WORLD
THEORY: HOW SOUND?

Fortunately, there are knowledgeable respected scientists who do not see a requirement to bend facts to fit a prevailing philosophy on the origin of life. Rather, they permit the facts to point to a reasonable conclusion. So, what facts and which conclusion? Interviewed in a documentary film, Professor Maciej Giertych, a noted geneticist from the Institute of Dendrology of the Polish Academy of Sciences, answered: "We have become aware of the massive information contained in the genes. There is no known way to science how that information can arise spontaneously. It requires an intelligence: It cannot arise from chance events. Just mixing letters does not produce words," He added: "For example, the very complex DNA, RNA, protein replacing system in the cell must have been perfect from the very start. If not, life systems could not exist. The only logical explanation is that this vast quantity of information came from an intelligence." Again, Professor Michael J. Behe stated: "To a person who does not feel obliged to restrict his search to unintelligent causes, the straight forward conclusion is that many biochemical systems were designed. They were designed not by laws of nature, not by chance and necessity; rather they were planned . . . Life on earth at its most fundamental level, in its most critical components, is the product of intelligent activity." Note; Professor Behe opens his statement: "To a person who does not feel obliged to restrict his search to unintelligent causes." Unfortunately young people venturing into further education feel very much obliged to follow the line of 'spontaneous generation.' Speaking on the matter to a young friend of my granddaughter some years past and who was about to enter Cambridge to study Micro Biology, I stated: "You will have to fall in with the ideas of your tutors and peers on the subject of evolution, or be ridiculed." After qualifying in that subject, he went on to study medicine at Oxford, qualifying in surgery. It is a great pity that generally, those renowned institutions of learning, although making claims to open-mindedness, still cling to Darwin's view of the subject, often seasoned with spontaneous generation theories. It is a proven fact that without the intriguing teamwork that exists between protein and nucleic acid molecules within a living cell, life could not exist. Yet clearly, it is constitutionally prudent to act with constraint on the subject until becoming a noted scientist, when doubts can be expressed with a degree of impunity!

Known as "the RNA world theory," it is claimed that RNA by itself was the first spark of life. Researchers in the 1980's discovered under laboratory tests, that RNA molecules could act as their own enzymes by snipping themselves in two and splicing themselves back together. It was thus speculated, RNA might have been the first self-replicating molecule. The theory goes on to further speculate, that in time, those RNA molecules learned to form cell membranes and finally, the RNA organism gave rise to DNA. "The apostles of the RNA world," writes Phil Cohen in 'New Scientist,' "believe that their theory should be taken, if not as gospel, then as the nearest thing to truth." Should we take that to mean, gospel = truth, 'nearest thing to truth' = speculative optimism? Not very convincing!

ARE WE DESIGNED?

S ceptics, observes Cohen, "argued that it was too great a leap from showing that two RNA molecules partook in a bit of self-mutilation in a test tube, to claiming that RNA was capable of running a cell single-handed and triggering the emergence of life on Earth." There are of course other problems to consider. Biologist Carl Woes holds that "the RNA world theory . . . is fatally flawed because it fails to explain where the energy came from to fuel the production of the first RNA molecules." Researchers have never located a piece of RNA which is able to replicate itself from scratch. Then there is the issue of, where did RNA come from in the first place? Although "the RNA world" theory appears in many textbooks, most of it says researcher Gary Olsen, "is speculative optimism."

One theory touched upon earlier and espoused by some scientists, is that our planet was seeded with life from outer space. That however, does not address the question, what originated life? Saying that life comes from outer space, notes science writer Boyce Rensberger, "merely changes the location of the mystery." The issue still remains, because it does not explain the origin of life but sidesteps it by relocating that origin to another system or galaxy. We have considered the significance of chance, design and organization but not looked at specific examples of life and so let us make a start with ourselves.

To quote the psalmist once again: " . . . in a fear inspiring way I am wonderfully made." Well how wonderfully are we made? No doubt when starting our activities each morning we glance in a mirror to check our appearance. It is likely there would not be enough time to be contemplative, but let us take a moment to marvel at what is involved in that simple glance: Although not vital to life, our eyes enable colour vision. Our ears are positioned to provide stereophonic hearing, which enables us to locate the direction from which sound originates both of which we may take for granted. A book for acoustic engineers, comments: "In considering the human hearing system in depth however, it is difficult to escape the conclusion that its intricate functions and structures indicate some beneficial hand in its design." Moving on to our nose which manifests marvellous design: Besides providing a means of taking in air, it contains millions of sense receptors, enabling us to discern some 10,000 nuances of smell. In enjoying a meal, another sense comes into play. Thousands of taste buds convey flavours to us, yet other receptors on our tongue help us to feel say, a harmful fish bone. Without those last senses alone, how many of us would be dedicated or disciplined enough to take into our system, the required chemicals to sustain our well-being, if eating were not made a pleasurable experience? Not just good design, but surely a degree of love in that design is made manifest! It is granted that some creatures have a keener night vision, more sensitive smell, or more acute hearing with a wider frequency response, but the balance of man's five senses certainly permit him to excel in many ways. We are of course able to benefit from those capacities and abilities because of the inter-reaction with our brain.

OUR SUPERCOMPUTER—
THE BRAIN!

Although animals have functioning brains, the human brain is in a class by itself, making us undeniably unique. The uniqueness relates to our having a meaningful, lasting life.

For years man's brain has been likened to a computer, yet recent discoveries show that the comparison falls far short. "How does one begin to comprehend the functioning of an organ with somewhere in the neighbourhood of 50 billion neurons with a million billion synapses (connections) and with an overall firing rate of perhaps 10 million billion times a second?" asked Dr. Richard M. Restak. His answer: "The performance of even the most advanced of the neutral-network computers . . . has about one ten-thousandth the mental capacity of a housefly." Consider not only in performance, how much a computer fails to measure up to the remarkably superior human brain. What man-made computer can repair itself, rewrite its program, or improve over years? When a computer system requires adjustment, a programmer must write and enter new coded instruction. Our brain does such work automatically, both in our early years of life and in old age. It would not be an exaggeration to state that the most advanced computers are very primitive compared to the brain. Scientists have called it "the most complicated structure known" and "the most complex object in the Universe"

When considering the super computer 'Deep Blue' the question arose: "Aren't we forced to conclude that 'Deep Blue' must have a mind?" Professor David Gelertner of Yale University replied: "no. Deep Blue is just a machine. It doesn't have a mind any more than a flower pot has a mind . . . Its chief meaning is this: that human beings are champion machine builders." He also pointed out the major difference between the brain and a computer when stating: "The brain is a machine that is capable of creating an 'I' Brains can summon mental worlds into being and computers cannot." He concluded: "The gap between human and [computer] is permanent and will never be closed. Machines will continue to make life easier, healthier, richer and more puzzling. Human beings will continue to care, ultimately about the same things they always have: about themselves, about one another and many of them, about God. On those terms, machines have never made a difference. And they never will."

Another comparison was drawn by Steven Pinker, director of the Centre for Cognitive Neuroscience at the Massachusetts Institute of Technology: "Today's computers are not even close to a 4-year old human in their ability to see, talk, move or use common sense. One reason of course, is sheer computing power. It has been estimated that the information processing capacity of even the most powerful supercomputer is equal to the nervous system of a snail—a tiny fraction of the power available to the supercomputer inside [our] Skull."

To say the least, our brain is a highly flexible biological mechanism. It can keep changing depending upon the way it is used—or abused.

Two main factors seem to be responsible for its development throughout our lifetime—what enters it through our senses and what we choose to think about. "No one suspected that the brain was as changeable as science now knows it to be," writes Pulitzer prize-winning author Ronald Kotulak. After interviewing more than three hundred, researchers he concluded: "The brain is not a static organ; it is constantly changing the mass of cell connections that are deeply affected by experience."—Inside the Brain. The brain is also affected by our thinking. Scientists have found the brains of those who remain mentally active have up to 40% more connections between nerve cells than do the brains of the mentally lazy. Neuroscientists conclude: "We have to use it or we lose it." There would seem to be a loss of brain cells as a person ages and advanced age can bring memory loss. Although the difference is much less than once believed: A 'National Geographic' report on the human brain said: "Older people . . . retain capacity to generate new connections and keep old ones via mental activity." There can be no disagreement with scripture here, because it encourages readers to 'be transformed by making their mind over' or to be "made new" through "accurate knowledge" taken into the mind. (Romans 12:2, Colossians 2:10)

How super is our computer?

S ince we shall shortly take a brief look at the design in animals. It might be appropriate to consider the wonders of our brain and its capacity. Brain scans prove that the frontal lobe becomes active when we think of a word or call up memories. In fact it plays a special part in, we being who we are. "The prefrontal cortex ... is most involved with elaboration of thought, intelligence, motivation and personality. It associates experiences necessary for the production of abstract ideas, judgement, persistence, planning, concern for others and conscience ... It is the elaboration of this region that sets human beings apart from other animals." (Mareb's Human Anatomy and Physiology) Eduardo Boncelli, director of research in molecular biology in Milan, commenting on the distinction between man and ape says: "the human brain is composed almost exclusively of the [cerebral] cortex. The brain of a chimpanzee, for example, also has a cortex, but in far smaller proportions. The cortex allows us humans to think, to remember, to imagine. Essentially, we are human beings by virtue of our cortex." Evidence of that distinction is certainly manifested in what humans have accomplished in fields such as mathematics, philosophy and justice, which primarily involve the prefrontal cortex. Professor Paul Davies reflected on the ability of the brain to handle the abstract field of mathematics. "Mathematics is not something that you find lying around in your back yard. It's produced by the human mind. Yet if we ask where mathematics work best, it is in areas like particle physics and astrophysics, areas of fundamental science which are very, very far removed from everyday affairs. It suggests to me that consciousness and our ability to do mathematics are no mere accident, no trivial detail and no insignificant by-product of evolution." (Are we Alone?)

So why do humans possess a large, flexible prefrontal cortex which contributes to higher mental functions, whereas in animals this area is rudimentary or nonexistent? The contrast is so great that biologist's who claim we evolved, speak of the "mysterious explosion in brain size." Professor of biology Richard F. Thompson, noting that extraordinary expansion, admits: "As yet we have no very clear understanding of why this happened." Could the reason lie in man's having been created with this peerless brain capacity?

The motor cortex contains billions of neurons that connect with our muscles. Its features contribute to our being far different from apes or other animals. The primary motor cortex gives us "(1) an exceptional capacity to use the hand, the fingers and the thumb to perform highly dextrous manual tasks and (2) use of the mouth, lips, tongue and facial muscles to talk."—Guyton's Textbook of Medical Physiology. Over half the motor cortex is devoted to the organs of communication, which helps to explain the unparalleled communication skills of humans. Some 100 muscles in the tongue, lips, jaw, throat and chest cooperate to produce countless sounds. To draw another comparison: One brain cell can direct 2,000 fibres of an athlete's calf muscle, but brain cells for the voice box may concentrate on only two or three muscle fibres, suggesting that our brain is specially equipped for communication. 'The Language Instinct' comments: "No mute tribe has ever been discovered and there is no record that a region has served as a 'cradle' of language from which it spread to previously language-less groups . . . The universality of complex language is a discovery that fills linguists with awe and is the first reason to suspect language is . . . the product of a special human instinct." The actual information required to pose the simple question: "How are you today?" is stored in a part of our brain's frontal lobe known as Broca's area, an area which some consider as the speech centre. Nobel laureate

Sir John Eccles wrote: "No area corresponding to the . . . speech area of Broca has been recognized in apes."

Even if some similar areas are found in animals, the fact remains, scientists cannot get apes to produce more than a few crude speech sounds. We though, can produce complicated language. To do so, we put words together according to the grammar of the language we speak. Broca's area helps us do that both when we speak and write. Of course we are unable to exercise the miracle of speech unless we speak at least one language with understanding of its word meaning. To do so involves another special part of our brain, known as Wernicke's area. Here billions of neurons discern the meaning of spoken or written words. Wernice's area helps us to make sense of statements and comprehend that which we hear or read; thus enabling our learning information and making a sensible response. The key then to human intelligence far surpassing that of apes, is our use of syntax.

Dr. William H. Calvin explains: "Wild chimpanzees use about three dozen vocalizations to convey about three dozen different meanings. They may repeat a sound to intensify its meaning, but they do not string together three sounds to add a new word to their vocabulary. We humans also use about three dozen vocalizations, called phonemes. Yet only their combinations have content: we string together meaningless sounds to make meaningful words." Dr. Calvin noted that "no one has yet explained", 'the leap from the animals' "one sound/one meaning" to our uniquely human capacity to use syntax. "Is only man, Homo sapiens, capable of communicating by language? Clearly the answer must depend on what is meant by 'language'—for all the higher animals certainly communicate with a great variety of signs, such as gestures, odours, calls, cries, songs and even the dance of the bees. Yet animals other than man do not appear to have a structured grammatical language. And animals do not, which may be highly significant, draw representative pictures. At best they only doodle."—Professors R.S. and D.H. Fouts.

There is even more to fluent speech. To illustrate: A verbal "Hello" is able to convey a host of meanings. Our tone of voice reflects whether we are happy, excited, bored, in a rush, sad or frightened and may reflect even degrees of those emotional states. Another area of our brain provides information for the emotional part of speech. Therefore, various parts of the brain come into play when communicating. Having to teach chimps nonverbal communication, Dr. David Premack concluded: "Human language is an embarrassment for evolutionary theory because it is vastly more powerful than one can account for." Why do humans have this marvellous skill to communicate thoughts and feelings, to inquire and to respond? The 'Encyclopaedia of Language and Linguistics' states "[Human] speech is special" and admits, "the search for precursors in animal communication does not help much in bridging the enormous gap that separates language and speech from nonhuman behaviours." Ludwig Koehler summarized the difference: "Human speech is a secret; it is a divine gift, a miracle." There is indeed a huge difference between an ape's use of signs and the complex language ability of children! Sir John Eccles observed an ability "exhibited even by 3-year-old children with their torrent of questions in their desire to understand their world." He added: "By contrast apes do not ask questions." Yes. Only humans form questions, including those with regard to the meaning of life. Professor A. Noam Chomsky notes: "Turning to the mind, we also find structures of marvellous intricacy. Language is a case in point, but not the only one. Think of the capacity to deal with abstract properties of the number system, [which seem] unique to humans." Unlike animals which mainly live and act upon present needs, humans are able to contemplate not only the past but plan for the future. Concerning the future of our universe, physicist Lawrence Krauss wrote: "We are emboldened to ask about things we may never see directly because we can ask them. Our children, or their children, will one day answer them. We are endowed with imagination." Animals have a degree of memory thus enabling them to find their way home or recall where food may be.

Humans have far greater, almost limitless memory. One scientist estimated that our brain can hold information to the extent that it "would fill some twenty million volumes, as in the world's largest libraries."

Some neuroscientists estimate that during an average life span, a person uses a mere 100th of 1% of the potential brain capacity. Our brain of course, is not just some vast storage place for information, like a computer. Professors of biology, Robert Ornstein and Richard F. Thompson wrote: "The ability of the human mind to learn—to store and recall information—is the most remarkable phenomenon in the biological universe. Everything that makes us human—language, thought, knowledge, Culture—is the result of this extraordinary capability." Moreover, we have a conscious mind, a statement which on the surface seems quite basic, but sums up something which unquestionably makes us exceptional. It has been described as "the elusive entity where intelligence, decision making, perception, awareness and the sense of self reside." Consciousness, says one definition, is "the perception of what passes in a man's own mind." Modern researchers have made great strides in understanding the physical makeup of the brain and some of the electrochemical processes that occur in it. They are also able to explain the circuitry and functions of an advanced computer. We though, are conscious and aware of our being, a computer is certainly not. As demonstrated, the difference between brain and computer is vast. Professor James Trefil observed: "What exactly, it means for a human being to be conscious . . . is the only major question that we don't even know how to ask." Consciousness is "one of the most profound mysteries of existence," observed by Dr. Davis Chalmers, "but knowledge of the brain alone may not get [scientists] to the bottom of it." One of the reasons for the mystery could be, scientists try to understand the brain by use of the brain, which may be likened to the use of a 'negative feedback' system in an electronic amplifier, a proportion of the output, of which, can be used to modify both the input strength and characteristic.

19

Two animals under scrutiny: Which bits came first?

Why do we have a brain with so much capacity, that we hardly test a fraction of it during a normal lifetime? The wonderful Designer of life had good reason to do so, as we shall discuss later. Because had Darwin's ideas with regards 'survival of the fittest' been correct, we would only have developed the minimum amount required to carry out our functions efficiently! But in the meantime, let us consider just two examples of the thousands of animals on Earth and marvel at their "exquisite design." Firstly, the bombardier beetle is noted for its unique defence mechanism. When threatened, the insect sprays boiling, foul-smelling liquid from its posterior, warding off spiders, birds and even frogs. This beetle is equipped with a pair of glands which open at the tip of its abdomen. Each of these has a reservoir which stores an acidic compound and hydrogen peroxide as well as a reaction chamber filled with enzymes dissolved in water. To protect itself, the insect is able to squeeze the solution from the reservoirs into the reaction chambers in order to trigger a chemical reaction.

The result produces noxious chemicals, water and steam—at a temperature of about 100 degrees Celsius—which are sprayed onto an attacker. Although the chambers are less than one millimetre in length, the beetle is able to change the speed, direction and consistency of its toxic spray to meet the requirement. Researchers have studied the beetle to learn how to develop more effective and ecologically-sound mist systems. It uses not just one-way valves to allow chemicals into the reaction chambers but also has a pressure relief valve to expel them. Engineers hope to use spray technology based on the bombardier beetle's system in car engines and fire extinguishers, as well as in medical drug delivery devices such as inhalers. Professor Andy McIntosh of the University of Leeds says: "Nobody had studied the beetle from a physics and engineering perspective as we did—and we didn't appreciate how much we would learn from it."

Scientists are also in awe of the Gecko. Its ability to scale smooth surfaces, even scampering across a smooth ceiling without becoming detached from it! How does this amazing little lizard do it? The Bible says the gecko "takes hold with its own hands." (Proverbs 30:28) The gecko's feet do indeed resemble hands and they grasp smooth surfaces with amazing agility. Each toe contains ridges which have thousands of hair-like protrusions. Each of those protrusions, in turn contain, hundreds of microscopic filaments. The intermolecular forces (Van Der Waals force) emanating from those filaments are sufficient to support the lizard's weight—scurrying inverted on a wet glass surface. Scientists are analyzing the protein which is secreted from muscles, giving these underwater creatures that ability. Researchers naturally wish to make adhesives with a performance like the gecko's feet. Among other uses, this could have "a variety of medical applications," notes Science News magazine, "from bandages that stay put when wet to a tape alternative to surgical sutures."

After that brief consideration of just those two creatures, it is pertinent to ask the question, were they designed for a purpose or did they just appear by chance? It is most difficult to understand the reasoning of evolutionists. They state creatures, in order to survive and increase in number, have found it necessary to develop and grow physical attributes, as though by conscious thought they had a choice in the matter, or succumb to either environment or predator. This over a period of time of course, which would suggest on that basis, it would have required the bombardier beetle, in order to be fully protected, to have its weapon system in place whilst awaiting its body. Quite an illogical progression of development! Related to the latter thinking is a statement made by Professor Armand Marie Leroi, who on BBC4 television, unashamedly claimed, to ensure its survival, the giraffe was forced to develop its neck in order to reach food high up in trees. Seemingly, a close biological relative to the giraffe is a sheep. Both ruminants and since sheep are so successful feeding at ground level it does seem, according to Professor Leroi's claim, the giraffe went to a lot of trouble growing its neck.

We must also remember that these theories, with no factual substance, claim that such development would have taken thousands of years, in which time the poor giraffe should have worked up a very good appetite!

It would be claimed it is essential for men to fly, if not for expediency, pleasure or business reasons, then for military purposes. Scientists also tell us; unless we reduce greenhouse gasses we together with all life on our planet will become extinct. The import of which was unfortunately demonstrated on a small scale following the 911 disaster. A ban on commercial air traffic in USA airspace produced a most significant improvement in atmospheric quality during that relatively short period. Is it then reasonable, using the premise of those who claim living creatures adapt to a requirement, to expect over a period of time, man to develop his own flight wings and propulsion unit? The foregoing question may seem rather flippant, particularly when the dynamics and intricacies of flight feathers alone are considered, a subject which in its self would occupy many written pages of fact. But is that scenario so preposterous when measured against those who put forward chance, accident, and necessity, in order to defend against challenges to Darwin's ideas?

OUR MIND UNDER CONSIDERATION.

In mention of design intricacies and because it is so far removed from that of animals, we return to a final consideration of the human mind. We last touched upon consciousness, which we all experience. For example, our vivid memory of past events, are not mere stored facts, like computer bits of information. We are able to reflect upon our experiences, draw lessons from them and use them to shape our future. We are able to consider several future scenarios and evaluate the possible outcome and effects of each. We have not only the capacity to analyze, but to create, appreciate and love. We can enjoy pleasant conversations about the past, present and future. We have ethical values about behaviour and can use them in making decisions which may or may not be of immediate benefit. We are attracted to beauty in art and morals. In our mind we are able to mould and refine ideas and guess how others will react if they are put into practice. Such factors produce an awareness which sets us apart from other life-forms on earth. When a dog, cat or bird looks at a mirror, it responds as if seeing another of its kind. When we do so, we are conscious of ourselves as beings with the capacities mentioned. We are able to reflect upon dilemmas such as: 'Why do some turtles live for one hundred and fifty years and some trees live for more than a thousand years and yet an intelligent human will make news if reaching one hundred years of age?

Dr. Richard Restak states: "The human brain and the human brain alone, has the capacity to step back, survey its own capacity for rewriting our own script and redefining ourselves in the world, is what distinguishes us from all other creatures in the world." Man's consciousness baffles some. The book 'Life Ascending,' whilst favouring a mere biological explanation admits: "When we ask how a process [evolution] that resembles a game of chance, with dreadful final penalties for the losers, could have generated such qualities as love of beauty, truth, compassion, freedom and above all, the expansiveness of the

Human spirit, we are perplexed. The more we ponder our spiritual resources, the more our wonder deepens." Very true and thus, we might round off our view of human uniqueness by evidence of our consciousness illustrating why many are convinced that there must be an intelligent Designer, a Creator, who cares for us. "Why do people pursue art so passionately?" asks Professor Michael Leyton in 'Symmetry, Causality, Mind.' He pointed out, "some might say, mental activity such as mathematics confers clear benefit on humans, but why art?" Leyton illustrated his point by stating: "People travel great distances to art exhibits and concerts. What inner sense is involved?" Similarly, people around the globe hang attractive pictures or paintings on the walls of office or home. Most of us like to listen to music at home or in our cars. Why? It is certainly not because music once contributed to the survival of the fittest! Leyton says: "Art is perhaps the most inexplicable phenomenon of the human species." We all know; enjoyment of art and beauty is part of what makes us feel "human." Often when we look at a mountain torrent shimmering in sunshine, stare at the dazzling diversity in a tropical rain forest, gaze at a palm-lined beach or admire the stars sprinkled across the black velvet sky, are we not in awe? And why do we have an innate craving for things which, in reality, contribute nothing or little to our survival? So, from where, do our aesthetic values originate? If we discount a maker who shaped those values at man's creation those questions lack satisfying answers. That is also true regarding beauty in morals. Is it not logical to conclude? If the universe and our being alive in it are accidental, our lives can have no lasting meaning. But, if our life in the universe results from design, there must be a meaning.

LIFE—IS THERE A PURPOSE
AND WHY SO SHORT?

A nother facet of human consciousness is our ability to consider the future. When asked whether humans have traits which distinguishes them from animals, Professor Richard Dawkins acknowledged man has indeed, unique qualities. After mentioning "the ability to plan ahead using conscious, imagined foresight," Dawkins added: "Short-term benefit has always been the only thing that counts in evolution; long-term benefit has never counted. It has never been possible for something to evolve in spite of being for the immediate short-term good of the individual. For the first time ever it's possible for at least some people to say, 'Forget about the fact you can make a short-term profit by chopping down this forest; what about the long-term benefit?' Now I think that's genuinely new and unique."

William H. Calvin notes: "Aside from hormonally triggered preparations for winter and mating, animals exhibit surprisingly little evidence of planning more than a few minutes ahead." By contrast, humans consider the future, even the distant future. Some scientists contemplate what may happen to the universe billions of years hence. The Bible says of humans:

"Even time indefinite [the Creator] has put in their heart." The Revised Standard Version renders it: "He has put eternity into man's mind." (Ecclesiastes 3:11) The mere fact that our just giving passing thought to the concepts of the infinity of space and time, harmonizes with the claim that a Creator has put "eternity into man's mind."

"Strangely enough," notes Professor C. Stephen Evans, "even in our most happy and treasured moments of love, we often feel something is missing. We find ourselves wanting more but not knowing what is the more we want. Religion is deeply rooted in human nature and experienced at every level of economic status and educational background." That summed up the research carried out by Professor Alistair Hardy and presented in 'The Spiritual Nature of Man.' He confirms that which numerous other studies have established—man is God conscious. While individuals may be atheists, whole nations are not. The book 'Is God the Only Reality?' observes: "The religious quest for meaning . . . is the common experience in every culture and every age since the emergence of humankind." From where does this seemingly inborn awareness of God come? If man were merely an accidental grouping of nucleic acid and protein molecules, why would those molecules develop a love of art and beauty, turn religious and contemplate eternity? John Polkinghorne of Cambridge University observed: "Theoretical physicist Paul Dirac discovered something called quantum field theory which is fundamental to our understanding of the physical world. I can't believe Dirac's ability to discover that theory or Einstein's ability to discover the general theory of relativity, is a sort of spin-off from our ancestors having to dodge sabre-toothed tigers, something much more profound, much more mysterious, is going on . . . When we look at the rational order and transparent beauty of the physical world, revealed through physical science, we see a world shot through with signs of mind. To a religious believer, it is the mind of the Creator that is being discerned in that way."—'Commonweal'

22

WHY ARE WE HERE?— SOME COMMON BELIEFS.

As noted in the preceding pages, modern scientific discoveries offer an abundance of convincing evidence supporting that both the universe and life on earth had a beginning. What though caused that beginning? After studying the available evidence, many conclude, there must be a First Cause. Nonetheless, they may shy away from attaching personality to that Cause. Such reluctance to speak in terms of a Creator; mirrors the attitude of some scientists. Albert Einstein for instance, was convinced the universe had a beginning and expressed his desire "to know how God created the world." And yet he did not admit a belief in a personal God; he spoke of a cosmic "religious feeling which knows no dogma and no God conceived in mans image." Similarly, Nobel laureate chemist Kenchi Fukui expressed belief in a great framework in the universe. He said "This great link and framework may be expressed in words such as 'Absolute' or God." But he called it an "idiosyncrasy of nature." Now that belief parallels much of the Eastern religious thinking! Many Orientals believe nature came into existence of itself. That idea is expressed in the Chinese characters for nature which literally mean "becomes by its-self" or "self-existing." Einstein believed his cosmic religious feeling was well expressed by Buddhism. Buddha held it was not important whether a Creator had a hand in bringing forth the universe and humans. Similarly, Shinto provides no explanation of how nature came to be, Shintoists believing the gods are spirits of the dead which may assimilate with nature. Interestingly, such thinking is not very far removed from that which was popular in ancient Greece.

The philosopher Epicurus (342-270 B.C.E.) is said to have believed that 'gods are too remote to do any more harm than good.' Holding; that man is a product of nature, probably through spontaneous generation and natural selection of the fittest. This demonstrating, that similar ideas of today are in no way modern. Alongside the Epicureans were Greek Stoics, giving nature the position of God. They felt, cooperating with natural laws was the supreme good. Clearly, those views have a marked similarity to those expressed today. Nevertheless, we should not dismiss all information from ancient Greece as quaint history. In the first century C.E. in the context of such belief, a noted teacher presented one of history's most significant speeches. The physician and historian Luke recorded that speech at chapter 17 in the book of Acts of Apostles. It should help us to form our view of the First Cause and see where we fit into the picture. That famous teacher, Paul was invited to a high court in Athens. Once there, he faced Epicureans and Stoics, who gave nature the position of God. In his opening remarks, Paul mentioned having seen in their city, an Alter inscribed "To an Unknown God" (A' Gno'stoi The'oi'). It is thought; Thomas Huxley alluded to this in coining the term "agnostic" which he applied to those holding "the ultimate cause (God) and the essential nature of things are unknown or unknowable." Is the Creator, as many have held, "Unknowable?" That is a misapplication of Paul's phrase; it misses his point. He was not claiming the Creator to be unknowable but rather, unknown to those Athenians. Paul did not have at hand as much scientific evidence for a Creator as we today. However, he had no doubt about there being a personal, intelligent Designer whose qualities should draw us to him, for he went on to say: "What you are unknowingly giving godly devotion to, this I am publishing to you. The God that made the world and all things in it, being as this One is, Lord of heaven and earth, does not dwell in hand-made temples, neither is he attended to by human hands as if he needed anything, because he himself gives to all persons life and breath and all things. And he made out of one man every nation of men, to dwell upon the entire surface of the earth."(Acts 17:23-26) Rather than suggesting God was unknowable, Paul emphasized, those who constructed the Athenian alter, together with many in his audience did not know God. Paul then urged them—and all who have read his speech since, to seek to know the Creator, for "He is not far off from each one of us." (Acts 17:27)

RECYCLING:
CAN WE MATCH NATURE?

We have examined various lines of evidence pointing to a Creator: The vast, intelligently organized universe, which clearly had a beginning. Another is life on earth, including the design manifested in our body cells. And yet another is our brain, with our associated awareness of self and our interest in the future. We should also look at two other examples of the handiwork in creation which touch us on a daily basis. In doing so, we must ask ourselves: What does this show us about the personality of the One who designed and provided it? Observation of that creation tells us much about the One who was responsible for it.

Paul, speaking in Asia Minor offered an example: "In the past generations [the Creator] permitted all the nations to go on in their ways, although, indeed, he did not leave himself without witness in that he did good, giving you rains from heaven and fruitful seasons, filling your hearts to the full with food and cheer." (Acts 14:16) Paul in that example showed how the Creator, in making those provisions, has born witness to His personality.

Efficiency in recycling waste products is a big issue with both, governments, environmentalists and scientists at this time. Can we learn anything from creation? Food for both man and animals results from intricate cycles—including the water cycle, the carbon cycle, the phosphors cycle and the nitrogen cycle. It is general knowledge that in the vital process of photosynthesis, plants use carbon dioxide and water as raw materials to produce sugars, using sunlight as the energy source. The oxygen released by plants during photosynthesis, can hardly be termed a "waste product" because as we all know, it is a wonderful by-product and it is absolutely essential that we take oxygen into our lungs, in order to metabolize food in our body. We will all have studied this process in basic science classes, but it does not make it any less vital and it is just the start.

Let us remind ourselves, phosphorus in our body cells and those of animals, is vital for a transfer of energy. Again, we obtain it from plants which absorb inorganic phosphates from the soil, converting them to organic phosphates. We consume plants containing phosphorus in these forms and use it for vital activities. Thereafter, the phosphorus returns to the soil in the form of body "wastes" which can again be absorbed by plants. We also need nitrogen, which is part of every protein and DNA molecule in our body and essential for life. Although about 78% of the air around us is composed of nitrogen, neither plants nor animals can absorb it in that form. Therefore, nitrogen in the air must be converted into other forms before it can be utilized by humans and animals. That fixation occurs in various ways. One of which is by lightning which transforms some nitrogen into an absorbable form, which reaches earth with rain. That naturally provided fertilizer, is taken up by plants which are consumed by humans and animals, providing nitrogen, which once again is returned to the soil as ammonium compounds, some of which convert back into nitrogen gas. Nitrogen fixation is also accomplished by bacteria which live in nodules on the roots of legumes. These bacteria; convert atmospheric nitrogen into substances which plants can use. By eating green vegetables we take in nitrogen which our body requires to produce proteins. Amazingly, species of legumes are to be found in tropical rain forests, deserts and even tundra. If an area is burned over, legumes are usually the first plants to re-colonize.

What marvellous recycling systems these are: Each of which, makes good use of the wastes from the other cycles. The energy required comes principally from our sun—a clean, endless and steady source. How that contrasts with human efforts at recycling resources! Even man-made products, termed 'environmentally friendly' may not contribute to a cleaner planet because of the complexity of human recycling systems. In this respect, 'US News & World Report' pointed out that products should be designed so that their high-value components can be easily recovered by recycling. That of course is exactly what we observe in the natural cycles. So what do those observations reveal about the Creator's forethought and wisdom? In order to help us see further, some of His qualities, we should consider one more system—that of our immune system, because it also involves bacteria.

24

BACTERIAL WONDERS AND
OUR IMMUNE SYSTEM.

uman interest in bacteria frequently focuses on their harmful effects and yet most are harmless to humans, many actually being beneficial. Indeed, they are of life-and-death importance. Before moving to the crucial role bacteria play, as an Engineer, I cannot move on without describing the wonders of the 'Bacterial Flagellum.' Even under a powerful microscope, it appears tiny and insignificant. It has been compared to a powerful outboard motor attached to a boat. What is the bacterial flagellum? There are different kinds of flagella, but the bacterial flagellum (Latin "whip") is probably the most studied. Attached to the cell wall of bacteria, the flagellum rotates, enabling the micro-organism to go forward, stop, move in reverse and change direction. It is estimated that half of all the known bacteria are equipped with variations of flagella. The DNA in the bacteria or micro-organism contains the "drawings" of the flagellum and its propulsion unit. The entire assembly consists of about forty proteins, which can be compared to the parts in a motor. Amazingly, it builds itself in only twenty minutes! The publication 'The Evolution Controversy' states: "The bacterial flagellum includes a rotary motor that spins around at speeds of 6,000 to 17,000 rpm. Even more remarkable, it can change direction in as little as a quarter turn, and then spin at 17,000 rpm. in the opposite direction." From a dynamics perspective, although the mass is extremely low, that is a feat of performance to stagger any design engineer. It contains bushings, a universal joint and a rotor. New Scientist magazine calls the bacterial flagellum "a prime example of a complex molecular system–an intricate nanomachine beyond the craft of any human engineer." Scientists are baffled by the fact that the tiny bacterial flagellum self-assembles in the exact order required for all forty parts to fit together properly and function correctly. Did it come about by chance or was it designed?

Bacteria, is necessary in our digestive tract. Some 400 species in our lower intestinal tract alone, they help to synthesize vitamin K and process wastes. Cows turn grass into milk with the help of bacteria which are also vital for fermentation. Yet, if they get where they do not belong in our body, up to two trillion white blood cells fight the bacteria which might otherwise harm us.

Daniel E. Koshland, Jr., editor of 'Science Magazine' explains: "The immune system is designed to recognize foreign invaders. To do so it generates in the order of 10^{11} different kinds of immunological receptors so that no matter what the shape or form of the foreign invader there will be some complimentary receptor to recognize it and effect its elimination." One type of cell which the body uses for that purpose is the Macrophage; the name means "big eater" which is a fitting description since it devours foreign substances in our blood. After eating an invading virus, the macrophage breaks it into small fragments, then displays some protein from that virus as a marker. That bit of marker protein serving as a red flag to our immune system, sounding the alarm that foreign organisms are on the loose inside us. If another cell in the immune system, the helper or T cell recognizes the virus protein, it exchanges chemical signals with the macrophage. These chemicals, are extraordinary proteins which have a bewildering array of functions, regulating and boosting our immune system's response to invasion. The process results in a vigorous fight against the specific type of virus. Thus our body usually manages to overcome infection. Actually much more is involved, although even such a brief description reveals the complexity of our immune system. Now then, how did we obtain this intricate mechanism? It is given to us free of charge, regardless of our family's financial or social standing! Compare that fact to the inequity in healthcare available to most people.

"For WHO, growing inequity is literally a matter of life or death, since the poor pay the price of social inequality with their health," wrote director general of WHO, Dr. Hiroshi Nakajima. One Sao Paulo slum dweller said: "For us good healthcare is like an item in a window display in a luxurious shopping mall. We can look at it, but it is beyond our reach." Millions of people around the globe feel the same way. So reminding ourselves of the comparison drawn earlier, we must see the sense of love, impartiality and justice displayed by the One responsible for our magnificent design, the One, who endowed us with our wonderful immune system.

Dark Matter and Dark Energy: The e nigma!

The previously discussed systems are just basic examples of the Creators handiwork, but they do not reveal him to be a real and intelligent person. Christendom's clergy, even less so than in Darwin's day, are unable to help when it comes to an explanation of creation. Their confusion is evident, some claiming part creation part evolution being responsible for life on earth. Others, denouncing the whole Genesis account as myth. Then of course there are those who completely lack an understanding of the scriptures, who claim the earth was created in six twenty-four hour days. Is it any wonder why people have strong misgivings when considering the Genesis account of creation? Often, thinking the clergy know best, people have been misled by those so-called teachers, who make a cursory inspection of the Bible, take out the bits that suite them as individuals and mix them with a few myths, pagan festivals and apostate practices which are far removed from Bible truth.

That; instead of studying the whole Bible, by use of science and logical reasoning in order to prove the truths and authenticity of its content.

A fine example was set by the conscientiousness of those in first-century Beroe'a (Acts 17:11). If from the outset we are able to establish that all things came into existence by creation, we must also accept that the Creator or Designer would not move outside or deviate from the physical laws he established from the beginning of the universe. Why so? We may ask: The answer is, 'simply because those laws work perfectly!' Scientists speak of dark energy and dark matter. Without going too deeply into the subject, it is a fact, as already mentioned, the universe is not just expanding but that expansion is accelerating! Scientists wanted to know what form of energy was causing that accelerating expansion. For one thing, it seemed to be working in opposition to gravity; and for another it was not predicted by current theories. Appropriately, this mysterious form of energy has been named 'dark energy' and it makes up nearly 75% of the universe. The other "dark" oddity; dark matter, was confirmed in the 1980's when astronomers examined various galaxies. These, as well as our own, appeared to be spinning too fast to hold together. Evidently then, in view of physical laws, some form of matter must be providing the necessary gravitational cohesion: But what kind of matter? Because scientists have no idea, they have called the stuff dark matter, since it does not absorb, emit, or reflect detectable amounts of radiation. How much dark matter is out there? Calculations indicate it could make up 22% or more of the universe. So consider this: According to current estimates, normal matter accounts for about 4% of the mass of the universe. The two big unknowns—dark matter and dark energy—appear to make up the balance. Thus, about 95% of the universe remains a complete mystery! Science is in search of answers, but all too often one set of answers leads to another layer of puzzles. A fact of which has already been quoted but worth repeating: The Bible reminds us with the profound statement recorded at Ecclesiastes 3:11 "Everything [God] has made pretty in its time. Even time indefinite he has put in their heart, **that mankind may never find out the work that the true God has made from start to the finish**."

MATTER–PRODUCING ENERGY.

"Who can say whence it all came and how it all happened?" We find that poem in "The Song of Creation." Composed in Sanskrit, over 3,000 years ago and part of the Rig-Veda; a Hindu holy book. The poet doubted that even the many Hindu gods could know "how creation happened" because "the gods themselves are [later] than creation." Writings from ancient Babylon and Egypt offer similar myths about the birth of their gods in a universe already in existence. A key point however, is that those myths could not say where the original universe came from. There is though one creation record which is different. That particular record to be found in the Bible; opens with the words: "In the beginning God created the heavens and the earth." (Genesis 1:1)

Reiterating the quote from the 1910 'The Encyclopaedia Britannica': "Matter can neither be created nor destroyed;" That thought was influenced by the 18th century scientist Antoine-Laurent Lavoisier. However, in 1945 a flaw was publicly exposed in Lavoisier's law by the detonation of a super critical mass of uranium at Hiroshima. During such an explosion, different types of matter form, but their combine mass was less than that of the original uranium. Some of the uranium mass of course being converted into energy. Another problem with Lavoisier's law; arose in 1952. With the detonation of; a thermonuclear device (hydrogen bomb) hydrogen atoms combined to form helium. The mass of the resulting helium though was less than that of the original hydrogen: Again, a proportion of the hydrogen mass having been converted into explosive energy: As proven, a small amount of matter represented by an enormous amount of energy. That link between matter and energy, introduced by Einstein, who had formulated the relationship more than forty years before, expressed by his $E=mc^2$. From that formulation, other scientists were able to explain how the sun has kept shining for billions of years. The continuous thermonuclear reaction within, converts about 564 million tons of hydrogen into 560 million tons of helium every second; which indicates that 4 million tons of matter are transformed into solar energy; a fraction of which reaches earth and sustains life. A number of research teams around the world today are experimenting in an attempt to emulate the sun's controlled fusion. Conversely, as stated in the opening pages, energy can also covert to matter, as in today's particle accelerators. One might ask, 'so what does this have to do with the Bible record of creation?' As already stated, the Bible is not a scientific textbook as such and yet, it has proven to be up-to-date and in harmony with scientific facts. From beginning to end, it points by name, to the; 'One' who created all the matter in the universe. I make reference once again to Isaiah 40:26; which clearly describes the relationship between energy and matter.

With the information so far presented, it cannot be unreasonable to continue on the subject of Creation from the biblical description: Because, after creating the heavens, things both invisible and visible, the Creator, together with his Master Worker (more on this later) focused on the earth. The method used to bring about its production can only be guessed at. Maybe the variety of chemical elements making up our planet could have been produced directly from God's transforming unlimited dynamic energy into matter, which physicists today, tell us is quite feasible. Or, as many scientists believe, the earth could have been formed out of matter ejected from a supernova. Then again it may have been a combination of those two methods and others which science has yet to unravel. Whatever the mechanism, the Creator is the dynamic source of the elements which make up our earth, including all minerals essential for keeping us alive.

GENESIS ACCOUNT:
FEASIBLE? FIRST FOUR DAYS.

I t will be appreciated from that which has already been considered in this article, founding the earth would have involved much more than supplying all the materials in the right proportions. Earth's size, its rotational speed and the distance from the sun, as well as its axial inclination not to mention, the almost circular shape of its orbit around the sun had to be just right. In fact, just exactly as they are! Clearly that same Creator, set in operation the natural cycles which make our planet fit to support an abundance of life and the only one to do so, to our knowledge.

I have used the term "today" a number of times and I am in no doubt that the context in which it has been used will be understood, in that it refers to a period of time. Not of course, a period of twenty-four hours. Its meaning will be obvious and yet the significance is most important in understanding the creative days referred to in Genesis. Some claim them to be literally 24-hour days. This would mean that the entire universe and life on earth was created in less than a week! Thankfully, it is easily discovered that the Bible does not teach that. The book of Genesis was written in Hebrew. In that language, "day" refers to a period of time. It can be a lengthy one or a literal day of twenty-four hours. Even in Genesis, all six days are spoken of collectively as one lengthy period. (Genesis 2:4; compare 2 Peter 3:8) The fact is, the Bible reveals that creative days or ages, encompass thousands of years. It must be appreciated; the first chapter of the Bible gives partial details of some vital steps which God took to prepare the earth for human enjoyment. The chapter does not give every detail; as we read it, we should not be put off if it omits particulars which ancient readers could not have comprehended anyway. For example, in writing that chapter, Moses did not report the function of microscopic algae or bacteria. Such forms of life only came into human view after the invention of the microscope in the 16th century. Nor did he specifically report on dinosaurs, the existence of which; was deduced from fossils in the 19th century. Instead, Moses was inspired to use words which could be understood by people of his day—but words which were accurate in all they said about earth's creation.

As stated, creative days lasted for thousands of years. A person can see this from what the Bible has to say about the seventh "day." The record of each of the first six "days" ends saying, "and there came to be evening and morning, a first day" and so on. Yet we will not find that term after the record of the seventh "day." In the first century C.E., some 4000 years downstream in history, the Bible referred to the seventh "day" as still continuing. (Hebrews 4:4-6) Therefore, the seventh "day was a period spanning thousands of years. And so we can logically conclude the same applies to creative "days."

The First and Fourth "Days"—It would seem the earth had been established in orbit around the sun and was a globe covered by water before the beginning of the six "days" or periods of special creative works. "There was darkness upon the surface of the watery deep." (Gen. 1:2) At that early point, something—perhaps a mixture of water vapour, other gasses and volcanic dust—must have prevented sunlight from reaching earth's surface.

It is described thus: "God proceeded to say, 'Let there be light' and gradually light came into existence," or reached the surface of earth.—Genesis 1:3 translation by J. W. Watts. The expression "gradually . . . came" accurately reflects a form of the Hebrew verb involved, denoting a progressive action which takes time to complete. Anyone who reads the Hebrew language can find this form some forty times in Genesis chapter 1 and it is a key to understanding that chapter. (The Hebrews counted their day as commencing in the evening and running until the following sunset). Also, that which was started in one period did not have to be fully completed when the next period began. To illustrate, light gradually began to appear on the first "day," and yet it was not until the fourth creative period that sun, moon and stars could have been discerned. Gen. 1:14-19.

The Second and Third "Days"—Before the Creator made dry land appear on the third "day," he lifted some of the waters. As a result the earth was surrounded by a blanket of water vapour. This physical condition was touched upon earlier at page three. The ancient record does not—and need not describe the mechanisms used. Instead, it focuses on the expanse between the upper and surface waters, which it calls the heavens. Even today people use that term for the atmosphere where birds and aircraft fly. (See Psalm 8:8.) In due course, God filled the atmospheric heavens, with a mix of gases vital for life. During the creative "days" however, the surface water subsided so that land appeared. Perhaps using geological forces which are still moving the plates of earth; God seems to have pushed ocean ridges up to form continents. That would have produced dry land above the surface and deep ocean valleys below, which oceanographers have now mapped and are eagerly studying. (Compare Psalm 104:8&9.) After dry ground had been formed, another marvellous development occurred. We read "God went on to say: 'Let the earth cause grass to shoot fourth, vegetation bearing seed, fruit trees yielding fruit according to their kinds, the seed of which is in it, upon the earth.' And it came to be so."—Gen 1:11. Although having touched upon photosynthesis and recognizing it is essential to plants, we did not consider chloroplasts, which are found in green plant cells, converting sunlight to energy. "These microscopic factories," explains the book 'Planet Earth,' "manufacture sugars and starches . . . No human has ever designed a factory more efficient, or whose products are more in demand than a chloroplast." Indeed later animal life would be dependent upon chloroplasts for survival. We all know that without green vegetation earth's atmosphere would be overly rich in carbon dioxide, resulting in our death from heat and lack of oxygen. Some specialists give astonishing explanations for the development of life dependent upon photosynthesis. For example, they say that when single-celled organisms in the water began to run out of food, "a few pioneering cells finally invented a solution. They arrived at photosynthesis." Interesting! Photosynthesis is so complex that scientists are still attempting to unravel its secrets. Is it reasonable, let alone logical, to believe self-reproducing photosynthetic life arose inexplicably and spontaneously?

Or is it more reasonable to believe it exists as the result of intelligent, purposeful creation, such as that described and reported on in the Genesis account? The appearance of new varieties of plant life may not have ended on the third creative "day." It could have continued on to the sixth "day," when the Creator "planted a garden in Eden" and "made to grow out of the ground every tree desirable to one's sight and good for food." (Gen. 2:8 and 9) As mentioned, the earth's atmosphere must have cleared on "day" four, so that more light from the sun and other "luminaries" reached planet Earth. (Gen. 1:14 and 15).

Days five and six.

The Fifth and Sixth "Days"—During the fifth creative "day," the Creator proceeded to fill the oceans and the atmosphere with a new form of life—"living souls"—distinct from vegetation. Biologists speak, among other things, of the plant kingdom and the animal kingdom and divide them into sub-classifications. The Hebrew word translated "soul" means "a breather." The Bible also says "living souls" have blood. Therefore, we may conclude that creatures having both a respiratory system and a circulatory system—the breathing denizens of the seas and heavens—began to appear in the fifth creative period. (Gen. 1:20; 9:3, 4) On the sixth "day," God gave more attention to the land. He created "domestic" animals and "wild" animals, these being meaningful designations at the time Moses penned the account. (Gen. 1:24) And so it was in this sixth creative period that land mammals were formed. What though about humans? The ancient record tells us, eventually the Creator chose to produce a truly unique form of life on earth. He said to his Master Worker: "let us make man in our image, according to our likeness and let them have in subjection the fish of the sea and the flying creatures of the heavens and the domestic animals and all the earth and every moving animal that is moving upon the earth." (Gen. 1:26) Man would therefore reflect the spiritual image of his Maker, displaying His qualities.

And man would be capable of taking in huge amounts of knowledge. Thus, humans could act with an intelligence surpassing that of any animal. Also, unlike the animals, man was made with a capacity to act according to his own free will, not being controlled mainly by instinct. In recent years, scientists have carried out extensive research on human genes. By comparing human genetic patterns around the earth, clear evidence has emerged which confirms, all humans have a common ancestor, a source of DNA for all people who have ever lived, which includes ourselves. More than twenty years ago in 1988, 'Newsweek' magazine presented those findings in a report entitled "The Search for Adam and Eve." Those studies were based on a type of mitochondrial DNA, genetic material passed on only by the female. Reports in 1995 about research on male DNA point to the same conclusion—"there was an ancestral 'Adam,' whose genetic material on the [Y] chromosome is common to every man now on earth," as 'Time' magazine put it. Another article states: "The next place to start looking for patterns of heredity was on the Y chromosomal 'Adam,' the ancestor of all men, which was determined in 1997. Two different research groups, led by Peter Underhill at Stanford University of Arizona, each announced that Y chromosomal 'Adam' had also lived in Africa many years ago. This is the most recent Common male ancestor to all men in the world."

Whether those findings are accurate in every detail or not, they illustrate, the history we find in Genesis is highly credible, being authored by the One who was on the scene at the time. It is also significant that researchers attempting to track back to "Eve" noted a sudden and dramatic depletion in population stating: "What it suggested was a genetic bottleneck—a period in human history when the population was so small that the genetic expressions of a single woman have an impact on all humans living on the planet today." It would be most interesting to discover how long after 2370 B.C.E. that indicated bottleneck took place! Because, that date, as the Bible reports, was the date when Noah his wife and three sons and their wives were the only survivors of the world-wide "Deluge." An accurate genealogy from Adam to Joseph can be found at Luke 2:23-38. Luke is widely accepted; being both a physician and an accurate historian.

What a climax it was when God assembled some of the elements of the earth to form His first human son, whom He named Adam! The historical account tells us that the Creator of the globe and all life on it, put the man He had made in the garden-area "to cultivate it and take care of it." (Gen. 2:15) At that time the Creator may have still been producing new animal kinds, The Bible says: God was forming from the ground every wild beast of the field and every flying creature of the heavens and He began bringing them to the man to see what he would call each one; and whatever the man would call it, each living soul, that was its name." (Gen 2:19) The Bible in no way suggests, the first man, Adam was merely a mythical figure. On the contrary, he was a real person—a, thinking, feeling human—who could find joy working in that paradise home. Every day he learned more about what his Creator had made and what that One was like—his qualities, his personality. Then after an unspecified period of time, God created the first woman, to be Adam's wife. Further, God added greater purpose to their lives with this meaningful assignment: "Be fruitful and become many and fill the earth and subdue it, and have in subjection the fish of the sea and the flying creatures of the heavens and every living creature that is moving upon earth." (Gen1:27 and 28) Nothing can change that declared purpose namely, that the whole earth should be turned into a paradise filled with happy humans living at peace with one another and with the animals.

The material universe including our planet and life on it, clearly testify to God's wisdom. Therefore, He could obviously foresee a possibility that in time, some humans might choose to act independently or rebelliously, despite his being the Creator and Life Giver. Such rebellion could disrupt the grand work of making a global paradise. The record says God set before Adam and Eve a simple test which would remind them of the need to be obedient. Disobedience, God said would result in their forfeiting the life which had been given them. The Creator was showing his care in alerting our first ancestors to an erroneous course, one which would affect the happiness of the whole human race. (Gen. 2:16 and 17)

GEOLOGICAL ACCURACY!

B y the close of the sixth "day," the Creator had accomplished everything necessary to fulfil his purpose. He could rightly declare everything he had made to be "very good." (Gen, 3:31) At this point the Bible introduces another important time period by saying that God "proceeded to rest on the seventh day from all the work that he had made." (Gen. 2:2) Since the Creator "does not tire out or get weary," (Isaiah 40:20) why is he described as resting? It indicates that He ceased performing works of physical creation; moreover, He rests in the knowledge that nothing, not even rebellion in heaven or on earth can thwart the fulfilment of His grand purpose. God confidently pronounced a blessing upon the seventh; creative "day." As a result, his loyal intelligent creatures—can be certain that by the end of the seventh day,"—peace and happiness will reign throughout the universe.

But, are we able to trust the Genesis account of creation and the prospects it holds out? As pointed out, modern genetic research is moving toward the conclusion stated in the Bible long ago. Also some scientists have taken note and compared the order of the creative progression presented in Genesis. Noted geologist Wallace Pratt commented: "If I as a geologist were called upon to explain briefly our modern ideas of the origin of the earth and the development of life on it to a simple, pastoral people, such as the tribes to whom the book of Genesis was addressed, I could hardly do better than follow rather closely much of the language of the first chapter of Genesis." He also observed: "The order as described in Genesis for the emergence of land, as well as for the appearance of marine life, birds and mammals, is in essence the sequence of the principle divisions of geological time." Viewing those statements compels us to consider: How was Moses—thousands of years ago—able to get the sequence of creative events correct if his source of information were not from the Designer and Creator himself?

"By faith," the Bible states, "we perceive that the universe was fashioned by the word of God, so that the visible came forth from the invisible." (Hebrews 11:3 'The New English Bible') Many are not disposed to accept that fact, preferring instead, to believe in chance or in some blind process which supposedly produced our universe and life. ★ But as we have seen, there are many varied reasons to believe that both the universe and terrestrial life—including that of our own—derives from an intelligent First Cause, a Creator; God.

★　For a study of the history of life forms on earth, see the book, Life-How Did It Get Here? By Evolution or by Creation?

The Bible frankly acknowledges "faith is not a possession of all people." 2 Thessalonians 3:2: However, faith is not credulity. Faith is based upon substance. Paul described it thus: "Faith is the assured expectation of things hoped for, the evident demonstration of realities though not beheld; For by means of this the men of old times had witness borne to them. By faith we perceive that the systems of things were put in order by God's word, so that what is beheld came to be out of things that do not appear." (Hebrews 11: 1-3)

THINGS WHICH ARE HIDDEN
FROM INTELLECTUAL ONES.

In conclusion, we have considered some scientific wonders of creation and although having alluded to the Creator by use of scriptural quotations, He has not been clearly identified by name, nor have we identified those he used to bring into being the universe and all which is in it. The Bible, from Genesis to Revelation is a logical compilation of explanation and instruction throughout. Nowhere does it disagree with scientific or archaeological discovery, its contents in fact are strengthened by up-to-date findings. There are of course, those who will go to great lengths to disagree with that statement, but never open minded enough to apply the same effort, perhaps because of a bias or long held views, to search for an understanding. But then understanding of such issues is not perceived with our brain only, but also with humility of heart. It is why the greatest man that ever lived upon the earth said: "I publicly praise you, Father, Lord of heaven and earth, because you have hidden these things from the wise and intellectual ones and revealed them to babes." (Matthew 11:25) In fact, in considering the complexities of the universe, all men should be as humble as babes. For centuries men have been convinced they know what is best for humankind and yet even with all our advances, having inherited a perfect planet, we are unable to organize ourselves to the extent required to provide for everyone on this planet or to eliminate war. Writing at Jerusalem in 1000 B.C., King Solomon was inspired to sum it up very well by stating, in all he had seen " . . . Man has dominated man to his injury." (Ecclesiastes 8:9) Evidence shows that statement to be as true today, as the day in which it was written.

Writings from ancient Babylon and Egypt, like those of the earlier referred to: Rig-Veda, offer myths about the birth of their gods, being in a universe already in existence. A key point here is that those myths were unable to say where the original universe originated. The Bible though, as we have seen, opens with the words: "In the beginning God created the heavens and the earth." Moses; was inspired to write that simple, dramatic statement some 3,500 years ago. It focuses on a Creator, a God who transcends the material universe because he made it and hence existed before it was. The Bible teaches "God is a Spirit," which means he exists in a form which we are unable to see. (John 4:24) Such an existence is more conceivable today, in that scientists have described powerful neutron stars and black holes in space, recognizing that invisible objects are detectable by the effects they produce.

Significantly, the Bible reports: "There are heavenly bodies and earthly bodies; but the glory of the heavenly bodies is one sort and that of the earthly bodies is a different sort." (Corinthians 15:40 and 44) In explanation, that does not refer to the invisible cosmic matter astronomers study. The "heavenly bodies" mentioned are intelligent spirit bodies. It may be wondered, who then besides the Creator, has a spirit body? In answer, according to the Biblical record, the visible realm was not the first thing created. This ancient account reports, the first step of creation was the bringing into existence of another spirit person, the firstborn Son. He was "the firstborn of all creation," or "the beginning of the creation by God." (Colossians 1:15; Revelation 3:14) He was also described as a "master worker" employed in all subsequent creative works. (Proverbs 8:22,30; see also Hebrews 1:1 and 2) This first created individual was unique. He was the only creation God produced directly and was endowed with great wisdom. Of him, the first-century teacher Paul wrote: "By means of him all other things were created in the heavens and upon the earth, the things visible and the things invisible." (Colossians 1:6; compare John 1:1-3) He was also referred to as the "Word," "the only begotten Son," "Messiah," "Teacher," "son of man." All those titles are descriptive and correct. So what are the invisible things in the heavens which the Creator brought into existence by means of this Son? Whilst astronomers report billions of stars and invisible black holes, here the Bible is referring to hundreds of millions of spirit creatures—with spirit bodies. Some may ask, 'why create such invisible, intelligent beings? Just as a study of the universe can answer some questions about its Cause, a study of the Bible can provide us with important information about its Author. For example, the Bible tells us he is "the happy God." Whose intentions and actions reflect love (1 Timothy 1:11 John 4:8) We can then, logically conclude, God chose to have the association of other intelligent spirit persons who could also enjoy life. Each, having satisfying work which was mutually beneficial and would contribute to the Creator's purpose: Since we are made in his image, it should be easy for us to understand the pleasure derived by being in the company of others. Nothing suggests those spirit creatures were to be like robots in obeying God. Rather, He endowed them with intelligence and free will. Again, Biblical accounts indicate that God encourages freedom of thought and freedom of action; confident that these pose no threat to peace and harmony in the universe.

The Son took on earthly form and became flesh. (John 1:4) Acknowledged historically by most, to be the greatest man that ever lived, as he was, Jesus Christ! For those who consider his conception by a virgin girl to be impossible, I ask, is it unreasonable to accept that the One, able to design and produce all life-forms including ourselves, would be unable to fertilise by spiritual means, a seed which any healthy human male is capable of physically doing? During his short life as a human upon earth, not only did he demonstrate the super-human power imparted by his Father, but he also made clear his belief in and understanding of the Hebrew Scriptures. Events, which many professing Christians, just as Darwin, through a lack of biblical understanding, cannot or will not accept.

31

WHO ARE RESPONSIBLE AND WHAT IS THEIR RELATIONSHIP?

This lack of understanding arises, mostly through the influence of apostate sectarian dogma or denominational religious teachings. Possibly the most important of those false teachings and incorporated into Christendom since the Nicaean Council of 325 C.E.; is the claim that Jesus is God. Never, did Jesus claim to be equal to his Father, because through his Father's power, he was commissioned to bring into existence the universe and all the things in it. His disciples once asked him how they might communicate with his father, the outcome from his reply, being what is known today as the model prayer or the 'our father.' It is found at Matthew 6:9-13 and opens: "Our father in the heavens let your **name** be sanctified." The importance of that request to God; cannot be over emphasized. Because God himself said to an obstinate Pharaoh: "But in fact, for this cause I have kept you in existence. For the sake of showing you my power and in order to have my **name** declared in all the earth." (Exodus 9:16) That of course has proven to be so because most have heard of those events recorded at Exodus but when we examine that "name" more closely we will appreciate that it is known and used throughout the earth quite regularly, by Christians and non Christians and often by comedians as the brunt of a joke. Whilst on earth, Jesus used the divine name on numerous occasions and in so doing, clearly shows the importance his Father attaches to its use. (John 17:26) We have all heard the saying: "What's in a name?" Well, if asked God's name, people will often answer "the Lord." Really though, that is no more informative in answering the question, "Who won the election?" by saying "the candidate." Neither of which, provide a clear answer, since "Lord" and "candidate" are not names. To illustrate further, a person may be called Sir, Boss, Dad, or Grandpa, dependent upon circumstance. Those titles may reveal something about the person, but the name of the person reminds us of everything we know about them. Likewise, titles such as Lord, Almighty, Father and Creator call attention to God's activities. But only his personal name reminds us of everything we know about him.

So then what is God's name? The oldest Hebrew manuscripts present the name in the form of four consonants, commonly called the 'Tetragrammaton' (from the Greek te.tra—meaning "four," and gram'ma "letter"). These four letters (written from right to left) may be translated YHWH (or JHVH). The Hebrew consonants of the name are therefore known. The question is, which vowels are to be combined with those consonants? Vowel points did not come into use in Hebrew until the second half of the first millennium C.E.★ Furthermore, because of a religious superstition which had begun centuries earlier, the vowel pointing found in Hebrew manuscripts does not provide the key for determining which vowels should appear in the divine name. That name occurs 6,828 times in the Hebrew text as (YHWH or JHVH). Although that is so, the greatest indignity modern translators render to the Divine Author of the Holy Scriptures is the removal or the concealing of his peculiar personal name.

★   Vowel points, a system of dots and dashes were devised by the Masoretes: Additionally, certain accent marks were supplied to indicate stress, pause, connection and clauses etc.

At some point a superstitious idea arose among Jews that it was wrong to even pronounce the divine name (represented by the Tetragrammaton). Just what basis was originally assigned for the discontinuance is not definitely known. Some hold, the name was viewed as being too sacred for imperfect people to speak. Another view being the intent was to prevent other non-Jewish peoples from knowing the name and possibly misusing it. Yet the Hebrew Scriptures themselves present no evidence that any of God's true servants ever felt any hesitancy about pronouncing his name. Non-Biblical Hebrew documents, such as the so-called Lachish Letters, show the name was used in regular correspondence in Palestine during the latter part of the seventh century B.C.E. The best known English pronunciation of God's personal name; is that of Jehovah. (Je. ho'vah) [the causative form, the imperfect state, of the Heb. Verb Ha, wah' (become); meaning "He causes to become"]. Although "Yahweh," is favoured by most Hebrew scholars.

Just as the reason or reasons originally advanced for the discontinuation of use of the divine name are uncertain, so too, there is much uncertainty as to when this superstition really took hold. Some claim it began following the Babylonian exile (607-537 B.C.E.). This theory however, is based upon a supposed reduction in use of the name by later writers of the Hebrew Scriptures, a view which does not hold up under scrutiny. Malachi for example, was evidently one of the last books of the Hebrew Scriptures written (in the latter half of the fifth century B.C.E.). And it gives great prominence to the divine name.

Many reference works have suggested that the name ceased to be used by about 300 B.C.E. Evidence for this date was supposedly found in the absence of the Tetragrammaton (or a transliteration of it) in the Greek Septuagint translation of the Hebrew Scriptures; begun about 280 B.C.E. It is true that the most complete manuscript copies of the Septuagint now known, do not follow the practice of substituting the Greek words Ky'ri.os (Lord) or The.os' (God) for the Tetragrammaton. These major manuscripts though, only date back as far as the fourth century C.E. More ancient copies, in fragmentary form, have been discovered, which prove the earliest copies of the Septuagint did contain the divine name. One of which is the fragmentary remains of a papyrus roll of a portion of Deuteronomy, listed as P. Fouad Inventory No. 266. It regularly presents the Tetragrammaton, written in square Hebrew characters, each instance of its appearance in the Hebrew text being translated. This papyrus is dated by scholars as belonging to the first century B.C.E. and thus, written four or five centuries before the earlier mentioned manuscript. A paper could be presented, which in itself offered overwhelming evidence to show the divine name was originally used regularly in both the Hebrew and Greek scriptures, giving proof of its substitution. Those scriptures tell us: "God is a spirit and those worshiping him must worship with spirit and truth." (John 4:24)

Since the first century C.E. apostasy has crept into Christendom, which has moved progressively further from the healthful teachings of scriptural truth. She has developed illicit, relationships with political factions and in so doing, has diluted Biblical truth to the extent that she has little or nothing to do with Christianity as Jesus taught it. In many instances she has promoted a belief in evolution and in some statements her clergy have claimed "God does not exist" or "is dead." The Bible predicting such a time; sums it up thus: "For there will be a period of time when they will not put up with the healthful teaching, but, in accord with their own desires, they will accumulate teachers for themselves to have their ears tickled; and they will turn their ears away from the truth, whereas they will be turned aside to false stories. (2 Timothy 4:3 and 4) At least the Scribes and Pharisees of Jesus' day, although described as hypocrites, understood why, how and through whom all things came into existence! But unfortunately, just like the clergy of today, they were very much involved with the political factions of their day.

32

CONCLUDING NOTES AND CREDITS

Unless otherwise stated, scriptural quotations have been taken from the 'New World Translation of the Holy Scriptures Revised A.D. 1961' and 'The Kingdom Interlinear Translation of the Greek Scriptures. 1969 C.E.: This, because of the recognized, reliability and accuracy of those sources. Not being influenced by sectarian dogma or denominational religious teachings, they impart greater understanding of the original text. Just one example being Christendom's voluntary confusion over the relationship between God the Father and His Son Jesus to be found at John 1:1. This scripture identifies Jesus (meaning: "Jehovah is Salvation") as the "Word." A direct interpretation from the koine' Greek scripture reads: "In beginning was the "Word" and the Word was toward the God and god was the Word." Clearly showing, as does all scripture; the true relationship between, The Son and his Father: "the God" and his Son as a "god." or "the only begotten-god." (John 1:18): Ignoring use of the 'definite article' in the Greek language, Christendom is unable to accept that relationship, illogically claiming both Jesus and God share equal status, together with a third being, known as holy spirit. Thus, worshiping a; "Trinity" of gods, as did the Babylonians, ancient Greeks and many other, pagan idol worshipers. The 'holy spirit,' as demonstrated throughout the Bible, being God's active force. (Compare Koehler and

Baumgartner's Lexicon in Veteris Testamenti Libros, Leiden 1958 pp. 877-879; Brown, Driver and Briggs' Hebrew and English Lexicon of the Old Testament, edited by G. Friedrich, translated by G. Bromiley, 1971, Vol. VI, pp. 332-451). The inspired Greek Scriptures were written, not in the ancient classical Greek nor in the modern Greek which dates from the fall of Constantinople in 1453 C.E., but in the common or Koine' Greek of the first century of our Common Era, which was the international language of the day.